Designing Environments

Series Editors

Michael U. Hensel, Head of Department of Digital Architecture and Planning, Vienna University of Technology, Vienna, Austria
Claudia R. Binder, Dean of EPFL ENAC, École polytechnique fédérale de Lausanne, Lausanne, Vaud, Switzerland
Defne Sunguroğlu Hensel, Associate Professor, School of Architecture, Southeast University, Nanjing, Jiangsu, China

This series seeks to address the unfolding climate, environmental and ecological crisis from a broad interdisciplinary perspective and in relation to the impact of human transformations of the environment. The aim is to shift away from segregated modifications of the environment divided into domains and scales, systems and objects, with at best minimum negative impact for the environment, towards an integrative interdisciplinary approach that understands, models and modifies the environment in comprehensive and integrative manner and with net positive impact on the environment. This endeavour involves earth, environmental and life sciences, environmental informatics, computer science, and the disciplines that centrally concern the transformation of the terrestrial environments, such as architecture, landscape architecture and urban design, as well as agriculture and food production. From a methodological perspective, computer and data science play a role in facilitating multi-domain and multi-scale models of environments with the purpose of both analysis and design. At the same time, the series will place the discussion in a necessary cultural context and also discuss the need for ethics in which an alternative approach to the transformation of the environment needs to be based on.

The Series Editors welcome book proposals on the following topics: paradigms, theory and methods for integrative inter- and transdisciplinary approaches to understanding, modelling and modifying environments; relevant historical and contemporary case studies; relevant current research projects; related data science and computer science approaches.

Federica Dal Falco
Editor

Beyond the Garden

Sustainable and Inclusive Green Urban Spaces

Springer

Editor
Federica Dal Falco
Department of Planning, Design,
and Technology of Architecture
Sapienza University of Rome
Rome, Italy

ISSN 2730-6526 ISSN 2730-6534 (electronic)
Designing Environments
ISBN 978-3-031-50566-9 ISBN 978-3-031-50567-6 (eBook)
https://doi.org/10.1007/978-3-031-50567-6

This Springer imprint is published by the registered company Springer Nature Switzerland AG
The registered company address is: Gewerbestrasse 11, 6330 Cham, Switzerland

Paper in this product is recyclable.

Acknowledgments

We would like to thank Alessandra Battisti for her contribution to the layout and structure of the book and for her precious references and advice. Thanks to Irene Poli for her careful and rigorous collaboration in collating the texts.

Contents

Chapter 1
Designing for the Post-anthropocene Era: The Garden as a Metaphor for a Holistic and Symbiotic Approach

Federica Dal Falco

Abstract The contribution introduces the theoretical approach consistent with the essays collected in this book, placing it in the framework of the activities of the Internationalization Unit of the Department of Planning, Design, and Technology of Architecture, of which the author is coordinator. Climate change issues are prioritized and closely linked to cultural and social aspects, considering biodiversity conservation in constructive dialogue with the needs of communities and the anthropical environment. The interdisciplinary and international character of the volume reflects the peculiarities and cultural project of the Department, also referring to the principles underlying the New European Bauhaus, the virtual creative space between art, science, and the humanities, recently launched by Ursula Von der Leyen. Consistent with this approach, the theme of the garden was chosen as an exemplary concept in the history of the relationship between nature, the artificial, and mankind. This vast and complex subject is examined from the theoretical and design viewpoints of various disciplines: environmental and technological design, botany, legislative aspects, behavioral psychology, history and criticism of art and design, architectural design, and urban and territorial planning. After analyzing the concept of the Anthropocene, the author proposes a reflection on the relationship between nature and the artificial, introducing the idea of gardens and landscapes as places of cultural acquisition. In the following paragraphs, a few examples of the extraordinary variety of green places are cited, divided into two macro-categories: the vegetable garden as an enclosed space, sacred, and physical at the same time, which has always been cultivated and guarded, and the garden as a mental place of theoretical and philosophical thought but also of play. These models constitute some of the references of the idea of the garden as a habitat that operates on several levels, integrating nature and culture. The last paragraph questions the theories and different approaches that consider the garden as a transversal device extended to the entire planet, from macro to micro. According to this approach, the conservation of biodiversity involves the

F. Dal Falco (✉)
Department of Planning, Design, and Technology of Architecture, Sapienza University of Rome, Rome, Italy
e-mail: Federica.dalfalco@uniroma1.it

F. Dal Falco (ed.), *Beyond the Garden*, Designing Environments,
https://doi.org/10.1007/978-3-031-50567-6_1

repositioning of humanity with respect to nature. Moving beyond the garden means to identify the design of green spaces as a cultural issue that implies a renewed awareness of the collaboration between the natural and artifact. In this sense, the garden is interpreted in its many dimensions and variations as a place of harmony and freedom, as an essential reference in the direction of overcoming anthropocentrism for a new narrative.

Keywords Anthropocene · Nature and artificial · Vegetable and urban garden · Italian garden and labyrinths · Planetary garden

1.1 Introduction: An Interdisciplinary Approach to the Theme of the Garden

Beyond the garden collects essays by professors and researchers who participated in, or later contributed to, the international study day organized in June 2022 by the Internationalization Unit of Planning Design Technology of architecture Department, Sapienza University of Rome. The Unit is an interdisciplinary study center with a focus on knowledge from urban planning, architectural technology, and design, pursuing a multidisciplinary and holistic approach through issues related to environmental, economic, and social sustainability, including knowledge in the fields of sciences and humanities. The objective of this dynamic and composite structure is to consolidate and expand internationally virtuous relationships and scientific and cultural collaborations with prestigious universities and research centers proposing an experimental model of comparison and sharing. In the research and dissemination activities carried out since 2020 and in relation to the book theme, the Unit has taken as its reference the principles that inspire the New European Bauhaus launched by Ursula Von der Leyen. This creative virtual meeting space is dedicated to innovative ways of living between art, culture, social inclusion, science, and technology. The new perspective operates under the sign of a multidimensional dimension to regenerate cities and territories through actions combined with ecological and digital transition, according to a concept of beauty based on values such as accessibility of spaces, preservation of biodiversity, and listening to diversity, in dialogue between the needs of communities and the human environment (Dal Falco et al. 2022, p. 280).

Consistent with this vision, Beyond the Garden offers an interdisciplinary contribution to this segment of green design that has always been exemplary in the history of the relationship between nature, the artificial, and humanity. Our approach encompasses theoretical research and case studies related to environmental design, botany, legislative aspects, behavioral psychology and human factors, art and design history and criticism, architecture, and urban and spatial planning.

The idea of the garden is investigated in a broad sense and from different angles, both as a polysemic concept and in its dimension as an experimental project.

The goal is to outline a framework of theories and best practices that envision new solutions in collaboration with nature, according to a systemic vision and renewed environmental awareness for a post-Anthropocene future.

Moving beyond the garden is not to deny the history of this place of mediation between environmental resources and human intervention. Nor does it negate its archetypal significance as an image of harmony and a metaphor for freedom.

On the contrary, I believe it is fundamental to reflect on the historical and documentary data that constitutes an indispensable reference for operating in an innovative project dimension. In this sense, the garden is considered as one of the main devices capable of rebalancing humanity's present and future relationship with nature, with the aim of safeguarding biodiversity and experimenting with new sustainable models.

1.2 Anthropocene: Climate Change Consciousness for Cultural Change

The Anthropocene is an Earth Sciences concept according to which the effects of human activity on the planet can be observed and measured. The new era has been defined by Paul J. Crutzen and Eugene F. Stoermer (Crutzen and Stoermer 2000; Crutzen 2002) and indicates the current geological period by identifying ecological decline and ecosystem collapse as the set of consequences produced by human civilization in every part of the globe. In 1988, Edward O. Wilson defined biodiversity as the richness and complexity of life on Earth while denouncing its alarming decline (Wilson and Peter 1988).

Compared to the 1980s, the loss of biodiversity is much more severe: glaciers are melting, sea levels are rising, and many species of animals and plants have disappeared or are on the verge of extinction (Boeuf 2014). The unsustainability of this impact has been scientifically proven by numerous studies. Such research has shown that man and technology are now the dominant geological actor and that political, social, and economic ecosystems are unequivocally intertwined with natural ones (Ellis 2018) (Fig. 1.1).

Environmental degradation has been defined as a phenomenon that causes less and less reversible damage for future generations and a profound impoverishment of the relationship between humanity and nature (Pearce 1983). This set of nefarious changes, which has accelerated dramatically over the past 50 years, necessitates a radical rethinking of the principles and rules pursued at the general and individual levels. Recently, the concept of the Anthropocene has been distinguished in its narrower meaning from the geological point of view and in its broader meaning related to the cultural sphere. But the same author (Trischlern 2016, p. 312) argued that there is no real division, because geological and cultural layers are closely related.

Fig. 1.1 Glaciers melt. 2017, Hoffell, Island. (Copyright: Federica Dal Falco (Author's photos))

The Anthropocene can thus be interpreted as a crisis inducing a cultural inversion, bringing about a break with the thinking of modernity and the idea that nature is only a source of resources for the development of human activities (Dal Falco et al. 2022, p. 275). Therefore, the new era can direct policy and individual behavior to cooperate with the planet by making sustainable choices from an interdisciplinary perspective. Within this framework, the garden is a transversal theme capable of promoting the culture of sustainability, identifying forms of coexistence between the reality of human needs, technological mediation, the environment, bodily experience, and sensory perception, looking to the future, but based on a tradition of specific studies that constitutes its cultural background.

1.3 Nature and the Artificial: Introduction to the Idea of the Garden

The issue of the Anthropocene in its cultural dimension leads to questions about the long-standing relationship between nature and the artificial. The prerequisite is an awareness of the importance that man-made objects have assumed with respect to the environment, propagating since the earliest times as an extension of the individual's operative faculties. The innumerable quantity of things that have come down to us coexists with what is churned out by industry in continuous evolution, with stratifications ranging from the archaeological to the contemporary, impossible to classify. It is a second nature, an artificial world that changes with us.

As Remo Bodei (2012) has argued, technologies have transformed *Homo sapiens* into the dominant living form on earth, with a progressive externalization of hand articulation and a search for ever more powerful and precise tools.

Over the centuries, the production of means of defense and offense has been fundamental on the strategic level of human evolution. But the consequence has been the progressive substitution of the environmental with the artificial (Natoli 2022, p. 32). This ability of humanity to adapt the environment to itself has increasingly divaricated the relationship with natural elements, to the point of perceiving them instead of as a co-inhabiting and at the same time antagonistic force, as something external. Vegetation and animals have gradually been extracted from their original and unified condition, forgetting that ours depends on their existence. The paradoxical distance between the planet and political and economic principles, values, and programs has led to the disavowal of the regenerative drive induced by nature, which is then our own culture.

As writer Paolo Volponi (1988) has argued, contemporary society has instrumentalized everything, keeping nature and the animal on a leash. It has reduced nature to a simulation keyboard where elements, such as the seasons, are the keys of this simulation plane (Volponi 1988, p. 156). But the planet is not a stereotypical landscape. In contemporary times, the evolutionary path that has allowed humans to place themselves at the center brings them back to the margin and forces them to redefine their place in the world (Natoli 2022, p. 13).

The Covid-19 pandemic has accelerated this process imposing the search for new compatible lifestyles. It is urgent and necessary precisely because of the intimate connections between nature and humanity. These two entities, which encapsulate harmony, dissonance, and transformation, constitute the essence of the entire earth and human reality, that is, nature itself. (Friedrich 1968, p. 330).

According to philosopher Salvatore Natoli (2022, p. 15), the ancient and ever-present compass for designing a responsible future lies in the ethics of virtue. This position is shared by chemist and environmentalist Enzo Tiezzi from a scientific point of view. Tiezzi and Marchettini (1999) argues that only awareness of the destructive capacity of human intervention combined with knowledge of sustainable development can lead to a rethinking of the proper relationship between nature and the artificial.

Even in the face of recent youth movements, this new sensibility signals an ecological awareness and the importance of its embedding in public policies and institutions (Smith and Bognar 2019).

On the other hand, the dimensions of people's bodies and operations are inseparable from any natural element, because we are tailor-made to coexist with our surroundings. Like the fruit of an orange. The designer Bruno Munari (1998) described the fruit by comparing it to an industrial product in its consistency between form, function, and consumption. Also, the anatomy of the citrus fruit responds perfectly to human anatomy. The shapes of the cloves are perfectly adapted to the arrangement of the teeth in the mouth, which easily draw out the juice when breaking them. So much so that tangerines could be considered a minor production, at the scale of children.

Thus, the tapestries of the *La Dame à la Licorne* cycle are preserved at the Musée de Cluny – Musée National du Moyen Âge in Paris. In the six scenes, the young woman is at the center of a circular meadow in the company of a lion, a handmaiden, and a unicorn, a sort of microcosm of pleasure and tranquility.

The artifacts were woven in Flanders between 1484 and 1500, in the *millefleurs* style, with a red background against which flowering branches surround the main figures, including dogs, rabbits, and a monkey. Five of the six pieces are interpreted as representing the five senses, in relation to their hypothetical proximity to spirituality, according to the hierarchical progression: touch, taste, smell, hearing, and sight. And in the tapestry dedicated to taste, an orange tree is noted, with its flowers and fruits (Erlande-Brandebourg 1978).

It is the theme of the garden. A place of amenity, productivity, meditation, and regeneration that from the earliest times has been a metaphor for the womb. This paradigmatic figure has represented the ethos of a unified cosmos where gods, plants, humans, and animals coexisted. From this archetype, garden art has always been a free play of natural elements, which over the centuries have taken on different forms and meanings. And it took shape in contemporary times as one of the most advanced laboratories of the preservation of the commons: air, water, greenery, and soil.

1.4 Landscape and Gardens as Places of Cultural Acquisition

The function of gardens is linked to the conception of landscape as a cultural acquisition. Over the centuries, these forms of human collaboration with nature have been transposed and transfigured into different visionary and symbolic dimensions that integrate knowledge, experience, and research in a wide variety of domains.

And the thought and artistic variations of different eras are expressed through plants, geometries, and plant compositions. Gardens are also part of the style of an era, where style means an autonomous reality that consists of a changing system of meaning-bearing forms, which together with the vision of the individual gives back that of a group in relation to the world (Shapiro 1953, p. 287).

The garden has always been linked to artistic images, literary descriptions, and philosophical theories. Thus, the landscape and the concepts of wild forest, river, mountain, or desert are filtered by oral, written, or visual narration.

As Assunto (1973) wrote, the discovery of a landscape involves the transformation of natural things into aesthetic objects, an operation carried out by the human being and his history. When the centuries-old manipulation of the environment presents symbiotic qualities, the landscape is memory (Shama 1997, p. 15). Consequently, every act that leads to its destruction involves the destruction of what has been done for it by culture and artistic civilization (Milani 2001, p. 40).

The landscape and the garden are complex objects. They are the result of a *double artialization*. This term indicates a process of artistic creation carried out in situ, through direct modifications of the soil and *in visu*, with works of works by painters, photographers, and writers, which intervene indirectly (Roger 1997, pp. 16–17). Their origin cannot therefore be traced back to the physical fact alone.

The site constitutes the zero degree, from which the landscape and garden are shaped through a metaphysical process. An example reported by the philosopher Alain Roger (1997, pp. 21–22) is that of the *Montagne Saint-Victoire*, now recognized as a landscape painted by Cézanne (Fig. 1.2).

The in *visu artialization* of the French painter gave the mountain a genius loci and a deep-rooted cultural stratification. An image so powerful that, following a fire, it proposed its restoration *à l'identique* of the paintings in which it is represented. An inversion that leads from the work *in visu* to the intervention in situ.

Even Yosemite Park, which from the second half of the nineteenth century was considered the first American Eden, would never have existed without the visits of preachers from the northeast and in particular without the activity of John Muir. He was one of the fathers of modern environmentalism and founded the Sierra Club, the first environmental protection association in the United States, in 1892. But above all, it was the photographs devoid of human traces taken by Carleton Watkins and

Fig. 1.2 Mont Sainte-Victoire (La Montagne Sainte-Victoire) (ca. 1892–1895) by Paul Cézanne. (Source: Public domain)

Ansel Adams and the paintings by Thomas Moran and Albert Bierstadt that convinced Abraham Lincoln and the American Congress to attribute national park status to Yosemite and Sequoia. As Simon Shama (1997, p. 10) writes, Yosemite's grassy valley floor was actually the result of logging fires set by the Ahwahneechee Indians, an impact of human activities on the earth's ecology, which in this case has produced a landscape of rare beauty.

The importance of landscape in the literary field was interpreted by Michael Jakob (2005) according to a diachronic reading that ranges from antiquity to Romanticism. It is a phenomenon that is difficult to define, as it is dual in itself, similar to a work of art and situated on the borderline between subjectivity and objectivity, freedom, and necessity (Jakob 2005, p. 7).

In particular, from the mid-eighteenth century, the transition took place from the description of nature to its representation with respect to the observer and therefore to the individual's experience in space. The external world is thus codified within a cultural framework, considering the perception and relationship that each person has with nature.

Jean Jacques Rousseau is among the proponents of this interpretation. The philosopher has in fact placed at the center of his reflections the fusional relationship between humanity and nature, elaborated by the feeling and conscience of the individual. In the book *Rêveries du promeneur solitaire* (2011), the philosopher contemplates nature in ten walks. He collects herbs and plants, reflecting on the alienation resulting from social conventions and material well-being and elaborates on what he considers to be the fundamental principle: the value of equality, expressed in the harmony between humanity and nature. Rousseau was buried on the Isle of Poplars, inside the garden of Ermenonville, where he lived his last years. The park was designed by Marquis René-Louis de Girardin, an admirer of Rousseau, and was one of the first English-style gardens in continental Europe.

In the following paragraphs, some examples of the extraordinary variety of green places are mentioned, attributable to two macro categories: the vegetable garden, a fenced space that is both sacred and physical, which has always been cultivated and guarded, and the garden, as a mental place of theoretical and philosophical thought but also of play. These models constitute some of the references to the idea of the garden as a habitat that operates on multiple levels, integrating nature and culture (Fig. 1.3).

1.5 *Horti*: Sacred, Paradisical, Healing

The garden is a centuries-old expression of human thought between theory, knowledge, and design. First of all, it is an idea whose forms and contents, from the small farmer's vegetable garden to the garden of the gods, bring together a vast range of vegetal objects, modeled using agriculture, hydraulics, architecture, and sculpture, for the purpose to create a natural environment in which to live and contemplate (Venturi Ferriolo 1988, pp. 14–15; Assunto 1988, p. 28).

Fig. 1.3 Yosemite Park poster. (Source: Public domain)

From ancient Persia to the gardens of Babylon, these welcoming places are associated with Paradise and the sacred, examples of harmony that arouse wonder and pleasant sensations.

The archetype of the landscape and the garden is Eden, as it was described in the second and third chapters of Genesis. Located in the East and watered by four rivers, it was inhabited by a man and a woman, by animals, and by every tree that was beautiful to look at and good to eat. Between pleasantness and usefulness, the first garden on the planet had the fatal tree of knowledge at its center, a sign of an initiatory journey between good and evil.

In Western culture, from medieval times, the forbidden plant was identified with an apple tree. The tree allowed man to acquire knowledge and learn about death, leading him to defend himself from the evils of nature connected to his precariousness (Natoli 2022, p. 35).

The image of the oasis is taken from the Koranic tradition with a different duality where the paradisiacal freshness contrasts with the scorching desert. Thus, the Islamic and Arab garden transposes its Eden into an urban dimension, carving it out in the ryad, enclosed by the walls of the buildings.

At the time of the Sassanids, the Persian culture developed a design structured into four symmetrical parts with a line of water that included a basin or pavilion in the center. Denial of extreme dryness, the garden creates a microclimate, becoming a model of regeneration through vegetation and water.

In its in *visu artialization*, the Persian garden is represented by the carpet, a concrete object and at the same time as light as an idea (Mascelloni 2013, p. 12).

This fascinating artifact conveys multiple narratives, from myths to the different perceptions that have intertwined on its very surface. The origin of this work of pure decoration dates back to nomadic cultures, and for the Oriental, it is the home in its primordial meaning (Riegl 1988). It is similar to a work of architecture, with the edge of the carpet enclosing the space, defining its shape.

The western *Hortus Conclusus* was protected by high walls and also derived from the idea of paradise. Thus, the medieval cloister is a square that is divided into crosses, with meadow areas, long-stemmed flowers, a well, and a fountain.

In the traditional spatial organization of the garden, the predominant geometric shapes are the square and the circle. The use of the circumference would indicate the need for the human mind to find its own center within itself.

From this typology arises the variant of the gardens of love, depicted in the miniatures of *Le Roman de la Rose* (1485) as reproductions of walled cities (Maresca 2004, p. 27). Architecture protects from nature and recreates it within the enclosure by attributing symbolic and alchemical meanings to the artificial and natural elements.

In his work on medicine *Paragrano*, Paracelsus (1493–1541) defines nature as the garden of knowledge, which provides the doctor with recipes for the body and the spirit (Paracelso 2022). It is the theme of healing plants, already described in Roman times in the lists of *Storia naturale* by Plinio il Vecchio (1997): they are medicines taken from vines, from olive trees and fruit trees, and from spontaneous trees and the medicinal herbs widely treated in books 25 and 27, with notes on magical rites to make them effective.

The study of plant properties for therapeutic purposes will be developed in the medieval monastic tradition, with the transformation of the garden into a place of secret cultivation, at the service of the so-called green alchemy. With Saint Hildegard of Bingen (1098–1179), women will have a prominent place in the research and practice of plant medical virtues, in harmony with nature and with a poetic dimension of existence (Maresca 2004, pp. 49–51).

Since the ninth century, the garden of the Benedictine abbess was chosen as a privileged place for the acquisition of such knowledge. But Hildegard had a pharmaco-botanical knowledge that went beyond the cultivations of her cloister (Moulinier, 1994, pp. 61–75). It is likely that the abbess was familiar with specialist literary sources, although she claimed that the descriptions were the result of direct observation and divine inspiration.

1.5.1 Botanical, Victory an Urban Garden

Research on medicinal herbs developed from the first half of the sixteenth century with university botanical gardens. The first examples were established in Pisa (1543), Padua (1545) and Florence (1545). In Florentine city, the green structure was called *Giardino dei Semplici*, where the *semplici* are plant varieties used as basic medicines (Ferri and Vannozzi 1993). These places of study and practice were built according to hermetic, alchemical, and astrological references, using diagrams based on the number eight. It is the symbol of regeneration, cosmic balance, and cyclical continuity.

In the garden of Padua, Daniele Barbaro had envisaged a square layout divided into four smaller squares, which were in turn divided into eight isosceles triangles. The whole was inscribed in a circle, which among the symbolic figures indicates the union of the sky with the earth. The garden of Pisa was founded by the botanist Luca Ghini by order of Cosimo I on a large rectangular plot. The field was divided into eight square flower beds organized according to complex symbolic designs, with the aim of capturing celestial energies (Maresca 2004, pp. 58–68) (Fig. 1.4).

In 1577, the Dutch botanical garden in Leiden was created, with exotic and unusual plant species in Europe, such as pineapples and tuberoses. The organization of the area involved the repetition of a module of four rectangles inscribed in squares, a high fence with one side occupied by a villa and a central pavilion. The rationality of the distribution corresponded to a precise idea of scientific classification of botany and to the care and observation of plant species (Vercelloni 1990, p. 52). Botanical gardens were imagined as collections of plants for teaching and research, but they were and are places that contain extraordinary intangible cultural heritage, small and large human narratives: travel, mysterious and exotic species, pioneering discoveries, and secret recipes.

In contemporary times, botanical gardens are defined as open-air green museums and custodians of plant biodiversity. They are increasingly opening up to a museum concept that also operates in the name of experimentation, creativity, and social inclusion (Viola and Speciale 2021, pp. 8–9).

Fig. 1.4 Botanical Garden of Padua. (Copyright: Federica Dal Falco (Author's digital elaboration))

Finally, there is the reference to the urban garden, a typology that has played a central role in times of crisis, for food supply, but also as a bucolic refuge in nature.

The spread of these typologies was one of the most tangible consequences of the war-induced food crisis. Victory gardens, also called war gardens or defense gardens, were established in the first decades of the 20th century. The cultivated lands occupied large or small public and private areas, located both in the centers and on the outskirts of the most important metropolises of the United States, Canada, Great Britain, also in Germany and Italy. The United States Department of Agriculture encouraged the planting of victory gardens during the course of World War II. Around one third of the vegetables produced by the United States came from Victory gardens (Kallen 2000) (Figs. 1.5 and 1.6).

In Britain, sports fields and golf courses were requisitioned for agricultural purposes. In 1943, most English families had their own vegetable gardens, and food cultivation was extended to much of the arable public land: in cultivated lands, in railway edges, and in meadows. Urban agriculture was supported by Dig for Victory, an initiative of the Ministry of Agriculture with dedicated advertising campaigns.

Leaflets, posters, and specific publications were widely distributed, where rational cultivation models were proposed and practical advice was provided on how to organize a vegetable garden, cultivate it, fertilize it, and keep it productive. On the advice of nutritionists, booklets were printed with recipes based on the wise and rational use of the few resources available. Among various documents exhibited in the *Gardening in times of crisis* section, as part of the *Garden Future. Designing with nature* at the Vitra Museum, there is a photograph that portrays one of the most emblematic Victory Gardens (Eisenbrand 2023, pp. 82–89). It is the image taken by the photojournalist Harry Shepherd in July 1942 of the crater of a Nazi bomb

Fig. 1.5 Publication of Royal Horticultural Society London. (Source: Public domain)

Fig. 1.6 ABC of victory gardens: backyard farming made easy for all. (Source: Public domain)

probably intended for Westminster Cathedral, which fell in a nearby square. The large circular void was transformed into a vegetable garden, bordered by walls made with bricks from the collapsed houses. The place became one of the symbols of the British resistance and also a saying "when the Nazis sowed death, a Londoner and his wife have sown life-giving vegetables" (Eisenbrand 2023, p. 85).

In Italy, the war garden was interpreted by the fascist regime in terms of food self-sufficiency to contribute to the autarchic policy. From 1940, the transformation of public gardens and parks in large cities into arable areas, where wheat, barley, Legumes, and potatoes could be planted and began in a systematic way. The cultivations were extended to the most important squares in Italy, such as Piazza Venezia.

The vegetable gardens of fascist Italy were supported by massive advertising propaganda and radio messages, inviting the people with imposing tones to cultivate even small flowerbeds and terraces, for the sustenance of their families.

Thus, while fascist ideology attributed to food gardening actions aimed at strengthening the identity of the regime, in the allied countries, the meaning of this structure promoted principles of resistance against the war unleashed by Nazi-fascism. In this sense, even a vegetable garden can become a political manifesto and ideological symbol.

Today, the relaunch of urban gardens recovers some values of the alternative and environmentalist community gardens of the counterculture in the sixties, contextualizing them in relation to current problems (Panzini 2021). The practice of self-managed cultivation has spread with a view to the regeneration of abandoned land, against building speculation and degradation, especially in the outskirts of cities with benefits for the health of citizens. Urban gardens contribute to making cities more ecological and livable places, improving the sustainability of the food chain.

Fig. 1.7 Victory garden, 1943. (Source: Public domain)

And once again, caring for the earth offers a broad-spectrum response, which combines environmental benefits with economic and social ones in terms of aggregation and inclusion (Fig. 1.7).

1.6 Gardens as Mental Places: Academic, Initiatory, Labyrinthine

Between the end of the fifteenth century and the beginning of the sixteenth century, the Italian garden developed as an evolution of the medieval type. The layout is geometric, with flowerbeds, evergreen hedges, and floral decorations on the lawn or on a colored gravel background. The garden included waterworks and sculptures with trees pruned and planted in groups and citrus fruits arranged in espaliers or in pots. The Medici Gardens were built in the Florentine countryside during the rule of the Medici family and are an exemplary reference in terms of the gardening and architecture of their villas. But, at the time, they were mainly considered places for philosophical reflection.

In particular, the Villa of Careggi and its garden were consecrated to the *otium philosophicum* and identified as places of verification of the theoretical speculations of the Florentine Platonic Academy. The villa is a work by Michelozzo and was donated in 1462 by Cosimo de' Medici to Marsilio Ficino, to allow him to develop studies on Plato. Believing that the stars influenced the correspondence between microcosm and macrocosm, Ficino also devoted himself to research into the therapeutic virtues of plants (Maresca 2004, p. 53).

The references to alchemical culture and Neoplatonic speculations are intertwined with a highly modern literary work, the *Hypnerotomachia Poliphili* (1499), which is attributed to Francesco Colonna. This singular allegorical novel narrates.

Polifilo's initiatory journey in search of his beloved Polia. The places described through texts and 169 woodcuts engraved on wood are largely inspired by the idea of a Renaissance garden. This model was pursued following the rediscovery of the ancient Roman garden and through the study of the classical texts that describe it. It represents the place of manifestations of the spirit and beauty, as well as the relationship between the ideal man and the perfect image of nature (Colonna 2010, pp. 1, 3).

Two centuries later, it is still the figure of the garden that represents a new school of thought, which proposes a return to classical themes. It is the Academy of Arcadia that developed and spread in Italy during the eighteenth century, in response to the Baroque considered a style of bad taste. It was founded in Rome on October 5, 1690, by Giovanni Vincenzo Gravina and Giovanni Mario Crescimbeni, with scholars linked to the circle of Queen Christina of Sweden.

The reference to the tradition of the shepherd-poets of Arcadia, and to its bucolic values, is also evident in the choice of the location and its name: *Bosco Parrasio*, a garden under the Janiculum, whose land had been donated by John V of Portugal. The architect Antonio Canevari transformed the impervious area into a garden by exploiting the slope of the terrain.

He built three floors connecting them with two flights of stairs alternately concave and convex, inserting an oval-shaped theater equipped with seats and lecterns. On the last difference in altitude, a building used as an archive and secretariat was erected, called the Reservoir (Fig. 1.8).

While the façade of the exedra dates back to 1838 and was designed by Giovanni Azzurri. Among laurels, magnolias, cypresses, pines, oleanders, and wisterias, the Arcadians recited verses, identifying the garden as the place of theoretical reflection and poetry linked to nature (Predieri 1990).

A theme profoundly connected to the idea of travel and the conception of the garden as an initiatory place is the labyrinth. This singular place, between play, confusion, and symbol of individual research, is defined as a winding path where it is easy to lose your way without guidance (Santarcangeli 1984, p. 27). It is above all a primordial idea, a system of signs that intersect with ethnology, archaeology, the history of religions, architecture, gardening, and poetry.

The labyrinthine structures have been imagined and built over the centuries under various forms and materials: from rock carvings, to indigenous drawings of the island of Malekula, to mosaic floors, to paths with plant species (Santarcangeli 1984, pp. 114, 178–185, 238–267). But the labyrinth is not always the work of man. There

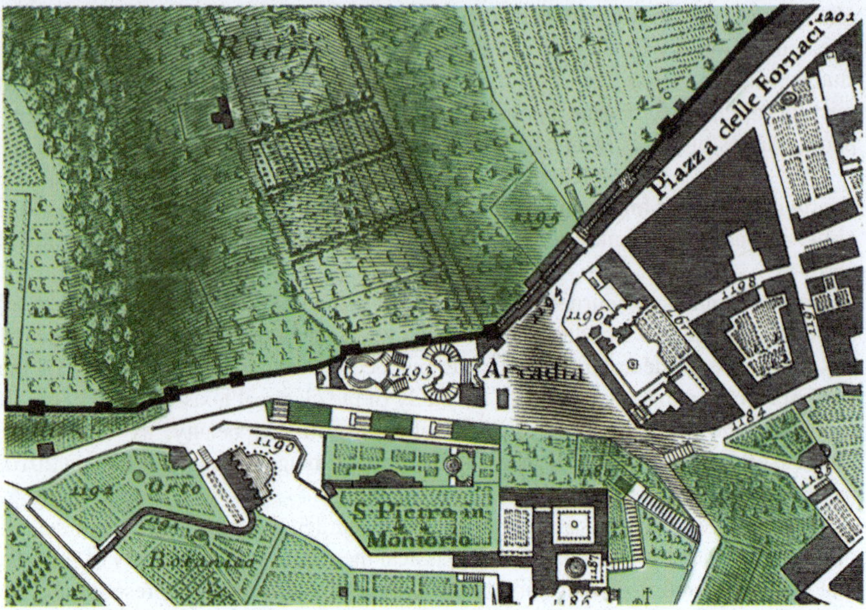

Fig. 1.8 Bosco Parrasio. (Source: Frutaz, A.M. (1962). (Ed.). *Le piante di Roma*. 3 volumi, Roma: Istituto di studi romani. Copyright: Federica Dal Falco. (Author's digital elaboration)

are natural structures, with systems of caves dug by the water, such as in Postojna, Slovenia, with complex networks extending up to 30 km.

It is no coincidence that the image is associated with the legend of Theseus, one of the most famous myths of antiquity. Famously, the labyrinth had been built by Daedalus in the Palace of Knossos, and the Athenian hero found the exit by unwinding the skein of thread that Ariane had given him (Fig. 1.9).

Then he killed the minotaur. According to Umberto Eco (1984, p. XIII), the labyrinthine structure can be traced back to three typologies: the unicursal, the manneristic, and the infinite network. The first involves an entrance and an exit, and it is intricate and difficult to navigate and reflects the idea of an equally complicated cosmos. But it has no false leads because it is orderly and designed. Defined as classical, it was found in rock carvings, the dating of which is conflicting. However, it's reasonable to assume that the symbol was known centuries before Christ in many parts of the globe and by different peoples.

The second model has a tree structure, with countless and complex ramifications, which often lead to a dead end, even though it has an outside and an inside. In the third type, this relationship does not exist, and each point can connect to another.

Between the fifteenth and sixteenth centuries, with the spread of gardening, the labyrinth was associated with the so-called topiary art of Roman origin. In addition to the yew, cypress, and boxwood, not only new world plants such as the gardenia and the orchid but also edible plants such as the tomato and the potato are introduced.

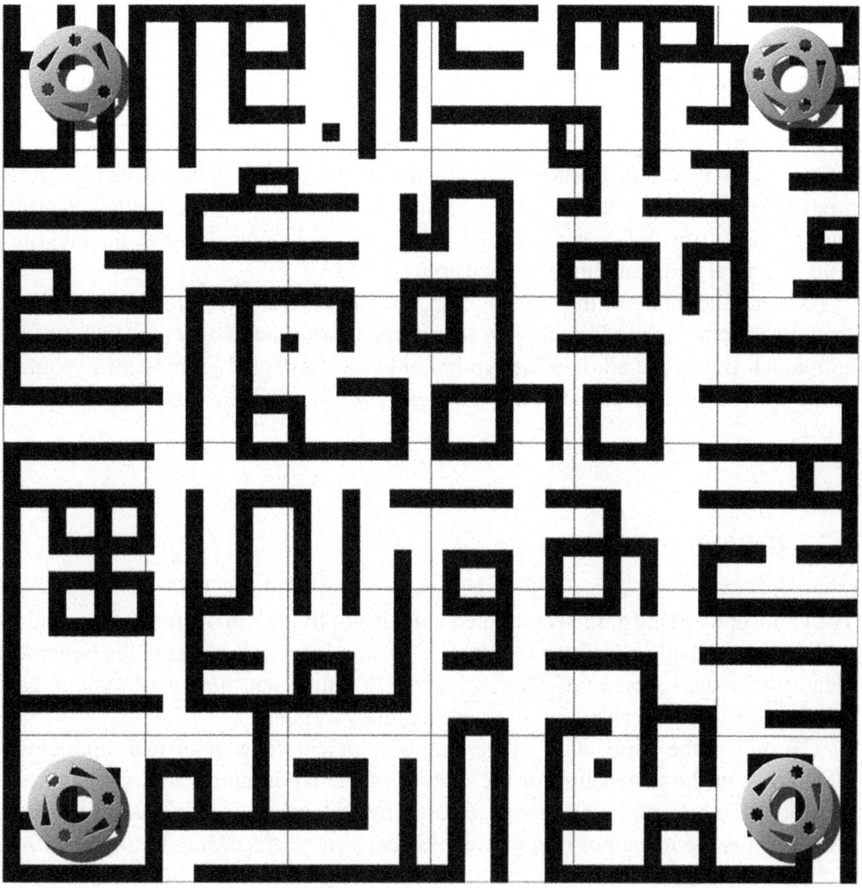

Fig. 1.9 Public space, 2017. (Source: Author's work. Copyright: Federica Dal Falco)

Humanistic culture used the word topiary as a transformation of plant forms into elementary geometries, according to an idea of mathematical domestication of nature (Vercelloni 1990, p. 35). The hedges of the labyrinths, artfully cut and all of the same height, are exemplary of the reduction of natural volumes to artificial forms, like the boxwood and yew labyrinths created in the gardens of the Venetian villas from the end of the fifteenth century, at Villa Pisani in Stra, at Villa Giusti in Verona, at Villa Barbarigo in Padua.

Later, the labyrinthine structures become more intricate and extravagant, until arriving at the *Labyrinthe de Versailles*, built by André Le Nôtre on the orders of Louis XIV. Initially, consisting of a vegetal grove, it was integrated with statues and fountains, which told Aesop's fables and constituting one of the most fantastic combinations of gardening, literature, sculpture, and hydraulic art.

In 2011, the Borges Labyrinth was created at the Giorgio Cini Foundation in Venice, designed by the architect Randoll Coate.

This homage to the Argentine writer is represented in the scrolls of the 3200 boxwood plants, through some of Borges' mental forms: the stick, the mirror, the hourglass, the question mark, and the tiger, integrated with the name Jorge Luis and the initials of the wife Maria Kodama.

In particular, the garden is inspired by the book *El jardín de senderos que se bifurcan* (1941) where Borges develops the theme of the ramification of various times, hypothesizing countless outcomes of an event, each of which entails a further multiplication of consequences. In this sense, the book coincides with the labyrinth in its temporal and emblematic dimension.

As symbolic figure of the human mind, the labyrinth is therefore the place where spatial and temporal coordinates change in relation to the experience of the single individual. But above all, it is the symbol of every existential journey, of remaining in the intricacies and then finding the right path in a game of mirrors of human events.

1.7 Beyond the Garden

The philosophy of the garden combined with that of living can overcome the contrast between nature and culture, since culture is a natural fact as it is part of the biological identity of *Homo sapiens*. Culture therefore as nature and not as its exploitation, without this meaning abandoning scientific innovation.

The aim is the application of the anthropological concept of anthropopoiesis, identifying in the acquisition of the culture of the environment and greenery, the possibility for humanity of a change in harmony with nature. And if nature is culture, the garden is the metaphor of a new ecological policy. According to Shama (1997, p. 537), they can be traced back to two types of arcadia: the wild one linked to Greek myth in a sort of continuity between men and animals and the orderly and harmonious bucolic one that becomes the reference for the country villas of Renaissance humanism. It is clear that the new approach is based on the awareness of climate change and the decline of postmodern thought, which placed the human factor at the center to the detriment of natural phenomena, often excluded from the narrative. Among the references on which the new paradigm is based, some radical ideas and experiences that date back to the mid-nineteenth century must be critically considered (Fig. 1.10).

Like the extreme positions of the founders of environmentalism, Henry David Thoreau and John Muir, based on the idea that the wilderness of the great American West was the antidote to the poisons of industrial society (Shama 1997, p. 7). In particular, Thoreau, with the book Walden or Life in the Woods (1988), offers the account of the author's experiment, who dedicated 2 years, 2 months, and 2 days (1845–1847) of his life in seeking an intimate relationship with nature, becoming a naturalist by chance.

The relationship with nature to counter capitalism and rapid industrialization was also at the center of the utopian vegetarian-nudist colony, founded in 1899 on Monte

Fig. 1.10 Monte Verità. From left to right: AW de Beauclair, Friederike de Beauclair, Hermann Hesse, Henry Oedenkoven, Ida Hofmann (maybe). Top right, Arnold Ehret. April 1907. (Source: Photo by FJ Hering. (Author's digital elaboration). Copyright: Federica Dal Falco)

Verità, near Ascona, in the Canton of Ticino. Between theosophy, free love, veganism, yoga, meditation, light baths, gardening, and the dances of Isadora Duncan, the international group of scholars and artists experimented with an outdoor lifestyle, living in huts and then transforming the community into a neo-avant-garde laboratory (Bollmann 2019).

In addition to the utopian movements of the twentieth century, including the radicals of the sixties, the recovery of the relationship with nature has been investigated by philosophers such as Leonardo Caffo. His theory of posthumanism questions what the new paradigm of life could be destined to replace *Homo sapiens*. The new figure of the posthuman should act in the awareness of being only a fraction of the multiplicity of organisms on the planet. And his ability to survive in an environment increasingly deprived of resources should rely on all his cognitive, behavioral, and relational characteristics (Caffo 2017).

The integration of the posthuman into nature offers a different narrative toward overcoming anthropocentrism. The prospect is to experiment with research paths and methodologies for a global and holistic design, the result of an interdisciplinary comparison. A design has been defined as symbiotic, with reference to the theories of botanist and plant neurobiologist Stefano Mancuso, according to which plants get the most from coexistence through plant relationships (Mancuso 2019).

Mancuso is an advocate of so-called nature-based solutions, such as intensive reforestation practices to reduce the concentration of carbon dioxide in the atmosphere. An idea that recalls the allegorical story by Jean Giono (1996), with the shepherd Elzéard Bouffier who manages to reforest an arid valley at the foot of the

French Alps on his own. It is also the conception of the philosopher Emanuele Coccia (2020) that since the world is mixture and relationship between things, its transformation can only be accomplished through its own components.

This approach finds its field of application in the contemporary design of urban green spaces, from vertical forests, to garden roofs of modernist origin, to the projects of French landscape architects (Racine 2006) and beyond, to the heterogeneous case studies exhibited in the *Garden Future.* In *Designing with nature* (Stappmanns et al. 2023, pp. 112–161), gardening is envisioned as an open work. A green laboratory where public space extends to the most disparate forgotten and abandoned places (Bourriaud 2018) from the largest areas to the smallest spaces, according to a porous and transformable interpretation of the boundaries capable of welcoming multiple concepts, hybrid forms, and lifestyles.

In its variants, even the humblest garden is therefore imagined as a transversal object extended to the entire planet, from the macro to the micro: from large extensions to interstitial ones. This is the theory of Gilles Clément (2005) who considers every place to be a place for the conservation of biodiversity.

To conclude, the garden not only is a concrete and naturalistic device but also is and has always been a place with multiple ethical and moral meanings: a model of life and freedom, an image, and metaphor of the coexistence of humanity, animals, and nature. Acting with respect for places means acting toward the planet and ultimately humanity, in the direction of overcoming anthropocentrism for a new narrative. And if culture is rethought within nature as a global phenomenon, it is possible to transform the garden into a world (Venturi Ferriolo 2019, p. 23).

This book proposes a critical contribution to these issues, questioning how to restore the totality of nature and culture by considering the Anthropocene as an opportunity, in search of a renewed balance between biodiversity and technological progress.

References

Assunto R (1973) Il paesaggio e l'estetica. Giannini, Napoli
Assunto R (1988) Ontologia e teleologia del giardino. Guerini, Milano
Bodei R (2012) Natura e artificio. Consorzio Festivalfilosofia, Modena
Boeuf G (2014) La biodiversité, de l'océan à la cité. Librarie Arthème Fayard et Collège de France, Paris
Bollmann S (2019) Monte Verità. 1900. Il primo sogno di una vita alternativa. EDT, Torino
Borges JL (1941) El jardín de senderos que se bifurcan. Sur, Buenos Aires
Bourriaud N (2018) L'Exforme. Art, Idéologie et rejet. PUF, Paris
Caffo L (2017) Fragile umanità. Il postumano contemporaneo. Einaudi Editore, Torino
Clément G (2005) Manifesto del terzo paesaggio. Quodlibet, Macerata
Coccia E (2020) Métamorphoses. Éditions Payot et Rivages, Paris
Colonna S (2010, May 14). La nascita dell'architettura del giardino rinascimentale nell'Hypnerotomachia Poliphili. BTA – Bollettino Telematico dell'Arte. http://www.bta.it/txt/a0/05/bta00562.html

Crutzen PJ, Stoermer EF (2000) The "Anthropocene". Glob Change Newsl 41:17–18
Crutzen PJ (2002) Geology of mankind. Nature 415(51):23
Dal Falco F, Veneziano R, Carlomagno M (2022) Natural and artificial interaction. Symbiotic processes between science, art and design. AGATHÓN Int J Arch Art Design 11:274–287
Eco U (1984) Prefazione. In: P. Santarcangeli, *Il libro dei labirinti. Storia di un mito e di un simbolo*. Frassinelli, Milano, pp XI–XIV
Eisenbrand J (2023) Gardening in times of crises. In: Stappmanns V, Kuijpers M, Heinrich M (eds) Garden futures: designing with nature. Exhibition catalogue. Vitra design museum, 25 march 2023–3 October 2023. Vitra Design Museum, Weil am Rhein, pp 82–89
Ellis EC (2018) Anthropocene. A very short introduction. Oxford University Press, Oxford
Erlande-Brandebourg A (1978) La Dame à la licorne. Éditions de la Réunion des Musées nationaux, Paris
Ferri S, Vannozzi F (eds) (1993) I giardini dei semplici e gli orti botanici della Toscana. Edizioni Quattroemme, Perugia
Friedrich H (1968) Montaigne. Gallimard, Paris
Frutaz AM (ed) (1962) Le piante di Roma, 3 volumi. Istituto di studi romani, Roma
Giono J (1996) L'homme qui plantait des arbres. Gallimard, Paris
Kallen SA (2000) The war at home. Lucent Books, San Diego
Mancuso S (2019) La nazione delle piante. Laterza, Roma Bari
Maresca P (2004) Giardini incantati, boschi sacri e architetture magiche. Angelo Pontecorboli editore, Firenze
Moulinier L (1994) Hildegarde de Bingen, les plantes médicinales et le jugement de la postérité: pour une mise en perspective. Scientiarum historia 20:61–75
Munari B (1998) Good design. Corraini Editore, Mantova
Mascelloni E (2013) Solo alla fine del mondo. In: Mascelloni E (ed) Tappeti Estremi. Da Timbuctù all'arte contemporanea, catalogo della mostra, Fondazione 107, 7 giugno – 10 novembre 2013. Edizioni 107, Torino, pp 12–38
Milani R (2001) L'arte del paesaggio. Il Mulino, Bologna
Natoli S (2022) Il posto dell'uomo nel mondo. Ordine naturale, disordine umano. Feltrinelli, Milano
Jakob M (2005) Paesaggio e letteratura. Casa Editrice Leo S. Olschki, Firenze
Panzini F (2021) Coltivare la città. Storia sociale degli orti urbani nel XX secolo. Deriveapprodi, Roma
Paracelso (2022) In: Masini F (ed) Paragrano. SE, Milano
Pearce D (1983) Ethics, irreversibility, future generations and the social rate of discount. Int J Environ Stud XXI(1):67–86
Plinio il Vecchio (1997) Storia naturale. Vol. 3–2: Botanica. Libri 20–27. Einaudi, Torino
Predieri D (1990) Bosco Parrasio: un giardino per l'Arcadia. STEM mucchi editore, Modena
Racine M (2006) Architectes de jardins et paysagistes de France. stiching kunstboek, Oostkamp
Riegl A (1988) *Antichi tappeti orientali*. Quodlibet, Macerata
Roger A (1997) Court traité du paysage. Éditions Gallimard, Paris
Rousseau JJ (2011) Rêveries du promeneur solitaire. Gallimard, Paris
Santarcangeli P (1984) Il libro dei labirinti. Storia di un mito e di un simbolo. Frassinelli, Milano
Shapiro M (1953) Style. In: Kroeber AL (ed) Anthropology today. An encyclopedic inventory. University of Chicago Press, Chicago, pp 287–312
Shama S (1997) Paesaggio e memoria. Mondadori Editore s.p.a, Milano
Smith EK, Bognar J (2019) A window for climate action. Social Science Open Access Repository. https://www.ssoar.info/ssoar/handle/document/65376
Stappmanns V, Kuijpers M, Heinrich M (eds) (2023) Garden futures: designing with nature. Garden futures: designing with nature. Exhibition catalogue. Vitra Design Museum, 25 March 2023–3 October 2023. Vitra Design Museum, Weil am Rhein, pp 112–161
Thoreau HD (1988) Walden ovvero Vita nei boschi. Rizzoli, Milano
Tiezzi E, Marchettini N (1999) Che cos'è lo sviluppo sostenibile? Le basi scientifiche della sostenibilità e i guasti del pensiero unico. Donzelli editore, Roma

Trischlern H (2016) The anthropocene. A challenge for the history of science, technology, and the environment. NTM Z Gesch Wiss Tech Med 24:309–335. https://link.springer.com/article/10.1007/s00048-016-0146-3
Venturi Ferriolo M (1988) Il giardino e l'esteta. Introduzione. In: R. Assunto, *Ontologia e teleologia del giardino*. Guerini, Milano, pp 9–15
Venturi Ferriolo M (2019) Oltre il giardino. Filosofia di paesaggio. Einaudi, Torino
Vercelloni V (1990) Atlante storico dell'idea di giardino europeo. Jaca Book, Milano
Viola A, Speciale M (2021) Andare per orti botanici. Il Mulino, Bologna
Volponi P (1988) Del naturale e dell'artificiale. Il lavoro editoriale, Ancona
Wilson, E. O. &, Peter, F. P. (1988). Biodiversity. National Academies Press

Part I
The City as an Open Laboratory of Interdisciplinary Knowledge and Practices Towards an Ecosystem Balance

Chapter 2
Avant Gardening to Grow the Green City

Alessandra Battisti and Livia Calcagni

Abstract Changing lifestyles and consumption-production patterns, as well as new highly efficient and soilless cultivation techniques, are evolving the way we design gardens within the city, making not only gardening but also farming possible in places where it was previously difficult or impossible. The garden can be conceived as a green infrastructure in the sense of a linked network of natural and seminatural elements capable of providing multiple functions and ecosystem services with positive economic and social benefits not only for humans. The new productive green city not only consumes less but is itself an autopoietic organism capable of addressing sustainability from not only an agrarian and architectural landscape but also a sociological, psychological, and educational point of view. Four case studies are used to argue whether verticalization is the best way to make cities greener, if the artificial reconfiguration of the environment can ever become a long-established practice, and what benefits does urban green lose when vertical solutions are adopted instead of encouraging direct soil cultivation directly on the ground level.

Keywords Green space · Green infrastructure · Garden · Urban farming

2.1 Introduction

Climate change and the progressive destruction of natural capital as a result of human activity, combined with migratory phenomena, wars, and pandemics, all exacerbated by the economic crisis (World Economic Forum 2023; IPBES 2019), force us to address the unavoidable environmental and social challenges. A wide range of geophysical conditions, including mean surface temperatures, ocean body temperatures, precipitation patterns, oxygen, and ocean acidity, are changing so quickly that nature cannot adapt to them, becoming no longer sustainable for many terrestrial ecosystems (Data-Driven EnviroLab & NewClimate Institute

A. Battisti (✉) · L. Calcagni
Department of Planning, Design, and Technology of Architecture, Sapienza University of Rome, Rome, Italy
e-mail: alessandra.battisti@uniroma1.it; livia.calcagni@uniroma1.it

F. Dal Falco (ed.), *Beyond the Garden*, Designing Environments,
https://doi.org/10.1007/978-3-031-50567-6_2

2020; European Commission 2015). Among these are ecosystem services necessary to provide goods such as food, fibers, fuels, water, and wood, as well as the regulation of environmental conditions such as pollution control, protection from natural hazards such as floods and forest fires, and water purification, which serve as the backdrop for major recreational, sporting, and cultural activities.

The need for action is urgent: 32% of the world's forests have been destroyed, 40% of invertebrate pollinators are at risk of extinction, and land degradation has resulted in a 23% decrease in land area productivity (WWF 2022; United Nations Convention 2022). The United Nations Global Agenda for Sustainable Development established the Sustainable Development Goals (SDGs) in 2015, indicating, among others, a series of objectives that specifically concern the territory and the soil, which had to be integrated into national short- and medium-term programs and were to be achieved by 2030 (Veerman et al. 2020):

- Ensure that land consumption does not exceed demographic growth.
- Ensure universal access to safe, inclusive, and accessible green and public spaces.
- Establish a land degradation-neutral world as a necessary prerequisite for preserving ecosystem functions and services.

By signing the Agenda, all European Union member states have chosen to participate in a process of monitoring these objectives, which is managed by the United Nations Statistical Commission and is based on the use of an indicator-based system, which includes a specific series of indicators concerning land take, land use, and the percentage of land subject to degradation (Munafò 2021).

2.2 Urban Green Transformation Scenarios

The risks posed by climate change and the alteration of environmental balances urgently call for the adoption of mitigation and adaptation measures (IPCC 2023), which in the urban context take on specific characteristics related not only to the anthropization of the territory, high population density, and mobility but also lifestyles and resource production (World Health Organization 2017). It has now been 7 years since COP21 in Paris, with only 7 years till 2030, the deadline for COP21 signatory nations to reduce emissions by 43% and adjust the track in terms of adaptation, funding, and losses and damages (Moosman et al. 2017). As stated by the United Nations, the world is lagging behind in achieving the Paris Goals, the effort made thus far is completely insufficient, and radical shifts in lifestyles and the way we inhabit the Earth are required if these goals are to be met in an increasingly short time frame (IPCC 2023).

According to sociologist Ulrich Beck, the future is a disease that forces populations to undergo more or less conscious transformations, and in his book *The Metamorphosis of the World*, he argues that in order to understand the metamorphoses of the world, one must investigate new beginnings, focusing on what is emerging from the old and attempting to glimpse future models and norms in the midst of the present turmoil (Beck 2016). The issues at the root of the change we are witnessing seem to be endemic: the city as the dominant form of living that keeps expanding and consuming land, uncontrolled human densities, vertical urban developments, and stratifications of endogenous social problems. It is urgent to address these issues to ensure not only the well-being of human habitats but also their survival (Brown and Aaron 2001; Battisti 2023). What instead has changed radically is the relationship between nature and artifice, as well as the awareness of the limit that characterizes it. Even before that, now that artificial intelligence can partly replace the human mind, the limit and control of technology has changed the very conception of the connections between the planet, humanity, and nature (Sadin 2018). Already at the end of the last century, the French sociologist Serres claimed in his book "The Natural Contract" that there is a need to establish a new contract that sanctions non-predatory ways of relating to nature and earth, recalling the need to rethink, in light of the natural contract, the profound experience of the sense of a religion (from the etymology of binding, connecting, reuniting) diligent rather than negligent of the world (since negligent is the opposite of religion). To put it in Serres's words "in our exclusively social contracts, we have dropped the bonds that attach us to the world, those that bind temporality to temperature, time to weather, those that put social sciences and physics, history and geography, law and nature, politics and physics, into relation; the bond that directs our language to silent, passive, obscure things that because of our excesses take back their voice, presence, activity, light. We can no longer neglect it" (Serres 1990).

Various tools have been developed at the European level to facilitate the achievement of the objectives of reducing greenhouse gas emissions, with a particular emphasis on decarbonization processes. Urban forestation is gradually spreading in Europe as a measure to increase ecosystem resilience and its capacity to absorb and compensate for external anthropic and natural pressures.

In Asian countries, a global race to build green cities seems to be underway, and this new green growth paradigm is leading to urban planning transformations and changes oriented toward an ecological version of urban entrepreneurship. Urban forestry and gardens are not primarily focused on solving environmental challenges but rather on redesigning urban ecology and the natural landscape at the service of economic competitiveness according to two different modes. The first is of a private, subjective nature and corresponds with the interpretation of forestation as an economic-financial investment project. The second has a public nature and concerns the production mechanisms of design response processes to promote urban health, leveraging the desire to return to living close to nature and promoting green areas with innovative experimental models of urban forestation resulting from the interaction of contemporary actors. In the second case, gardens and urban green spaces reflect identities, dreams, and visions of a near future that does not belong only to the economic or technical-administrative chain. The network must be expanded to include political, institutional, and social actors representing, in various capacities, the socio-technical context within which the green transformation production process takes place.

The subjects are responsible for the final consistency of the work as the tangible output of a culturally embedded manufacturing process. This approach renews and improves the garden, transforming it from a romantic idyll into a field of technological-environmental experimentation for the fight against climate change, for biodiversity promotion, and for the realization of social justice principles, transforming gardens into avantgarde places capable of mediating urban metamorphoses through processes of collective action, the comprehension of which is problematic but oriented straight ahead to a truly sustainable future.

2.3 Avant Gardening

The garden, as an idealized space that pervades our daily lives and imaginations, is a place where practical function and symbolic meaning merge and where the design and construction of experimental spaces can implement advantageous urban policies ranging from didactic-educational lessons to the creation of real financial modeling tools, all based on best practices and experimental data provided during the last few decades. We are witnessing design processes that involve expanding networks of financial partners and equipment linking project developers with innovative and evolutionary technologies for the production of ornamental and agricultural plants. There are several potential urban gardening possibilities that become sites where social justice and public involvement are disclosed and negotiated, ranging from social to political and financial interests and related to cultural value systems. As urban densities rise and land availability decreases, green spaces within cities tend to find space vertically: vertical gardens, farms and forests, rooftop gardens, and elevated structures for urban agriculture. These projects demonstrate the wide range of plant cultivation techniques—whether for human use or not—in different ways than the traditional ground-zero soil cultivation practice. Although vertical greenery can take many forms depending on the structure, the species, and the context, all of these solutions have one thing in common: they optimize open green areas while consuming as little horizontal space as possible. In this sense, the already widely recognized impact of plants on cities takes on a new dimension as a result of verticality. These alternative farming methods are used to reduce heat islands, increase biodiversity, promote food security, improve air quality, and reduce CO2 emissions, among other benefits.

2.4 Visionary and Unusual Avant Gardens: Blurring the Lines Between the Natural, the Artificial, and the Cultivated

Some questions arise spontaneously, and many wonder if verticalization is the best way to make cities greener and if the artificial reconfiguration of the environment can ever become a long-established practice. What are the long-term consequences of these activities in cities? Furthermore, what benefits does urban green lose when vertical solutions are adopted instead of encouraging direct soil cultivation directly on the ground level? The garden can be conceived as a green infrastructure in the sense of a linked network of natural and seminatural elements capable of providing multiple functions and ecosystem services with positive economic and social benefits for humans and other species (Benedict and McMahon 2012; Pitman et al. 2015; Naumann et al. 2011; Williamson 2003).

Blurring the lines between the natural, the artificial, and the cultivated, four very different projects are proposed as visionary and unusual *avant gardening* concepts that reflect the shifting mindset and cultures of their specific time and space. The case studies are thoroughly investigated in order to argue the advantages and disadvantages of vertical greenery in its broader sense. The projects have been analyzed according to different perspectives (location, spatial arrangement, functional purpose, and tenure) and to their compliance with the following parameters: minimum land footprint, microclimate improvement, biodiversity increase, CO_2 reduction, resource circularity, food security promotion, and social catalyst role.

2.4.1 Shiseido Forest Valley

The Shiseido Forest Valley (Fig. 2.1) is one of Asia's largest indoor gardens, spanning five stories and approximately 22,000 m^2 located in the heart of Jewel Changi Airport. It houses around 3000 trees and 60,000 shrubs of 120 species that live in high-altitude tropical forests from around the world and a 40-m-high indoor rainwater waterfall running through the middle of the garden that can channel 10,000 gallons of water a minute. The greenhouse houses shops and public spaces. The Airport is re-imagined as a major public realm attraction and vibrant urban center, which combines an intense marketplace and iconic gardens to create a new community-centric typology hub (Doe 2000). The vertical indoor garden offers travelers and visitors perfect microclimate conditions, unusual green and fresh spatial experiences, and contact with nature within the city (Safdie Architects 2023; Raut 2019). The indoor garden is not really open to the public, as visitors must pay an entrance ticket for most of the attractions, like the Canopy Park and Bridge, the Hedge Maze, and Foggy Bowls. Thus, it cannot be entirely considered a public green space open to the citizens. Moreover, although irrigation is provided thanks to rainwater collection, maintenance costs, and energy consumption are extremely high. In fact, indoor plants depend on artificial conditions for light, temperature, and humidity, making them more dependent on human intervention compared to outdoor gardens.

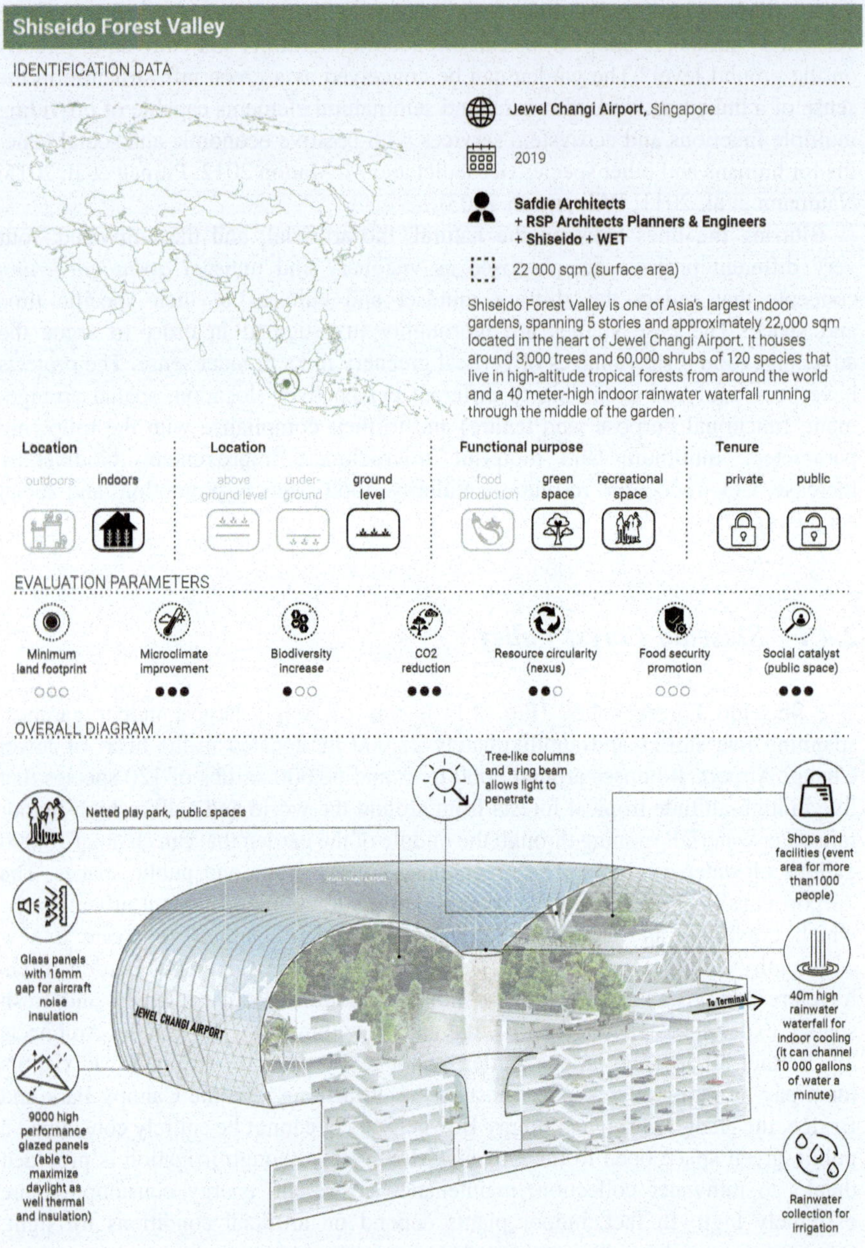

Fig. 2.1 Shiseido Forest Valley in Singapore Jewel Changi Airport with an indoor garden spanning five stories and a waterfall. (Source: Authors' elaboration)

2.4.2 Growing Underground

Another extremely interesting indoor garden is Growing Underground (Fig. 2.2), an underground farm located in London's former air raid shelters. In this case, urban gardening takes the form of urban agriculture using controlled environment techniques and hydroponics as an alternative local source of food. Integrating urban farms within dense environments using unused spaces, hardly usable for any other purpose, has the potential to use waste infrastructure and resources within cities and at the same time provide jobs and improve food security. The hydroponic system does not make use of pesticides and recirculates water (Chutiphongdech 2020). In order to optimize the use of resources and energy, a network of 25 sensors is installed in order to constantly monitor temperature, relative humidity, CO_2 concentration, air velocity, light levels, and several other parameters, which strongly affect the crops' growth. The sensors transmit data to loggers that are configured to process the data, and a web platform (digital twin) is used to visualize in real time the data acquisition (Walsh 2021; Jans-Singh et al. 2019). Moving production underground frees up more space on the surface to host a growing population and at the same time gives a new function to abandoned unused spaces. But in social terms, a subterranean garden doesn't provide a pleasant space for social interaction and engagement and has limited accessibility and nonoptimal comfort conditions for users.

2.4.3 Soradofarms

Reusing urban voids, leftover spaces, or empty unused spaces to provide the city with green infrastructure is a widespread strategy. Unused spaces are not only underground tunnels, as in the Growing Underground project, but also empty parking lots or small plots between buildings or even rooftops. Rooftop gardens and farms have been established all over the world to enable growing food in dense urban areas. In Japan, a whole new kind of urban rooftop farm was opened recently. Soradofarm (Fig. 2.3) is an urban agriculture project that uses the rooftops of train stations to accommodate urban gardens for commuters that want to use their transfer time to relax, regain connection with nature, and train their gardening skills (de Boer 2014). At the five rooftop allotments of Soradofarm, commuters can jump out of their busy lifestyles to grow fresh food. The train station rooftop farms act as urban catalysts, providing a multifunctional outdoor garden where commuters and their families as well as locals who have no space for a private garden can enjoy open-air time while engaging in community activities like farming, playing, or even having a picnic (Taramanni 2015). Moreover, the decentralized reconversion of concrete rooftops in urban community gardens contributes to the environmental maintenance and revitalization of the areas along the railway line.

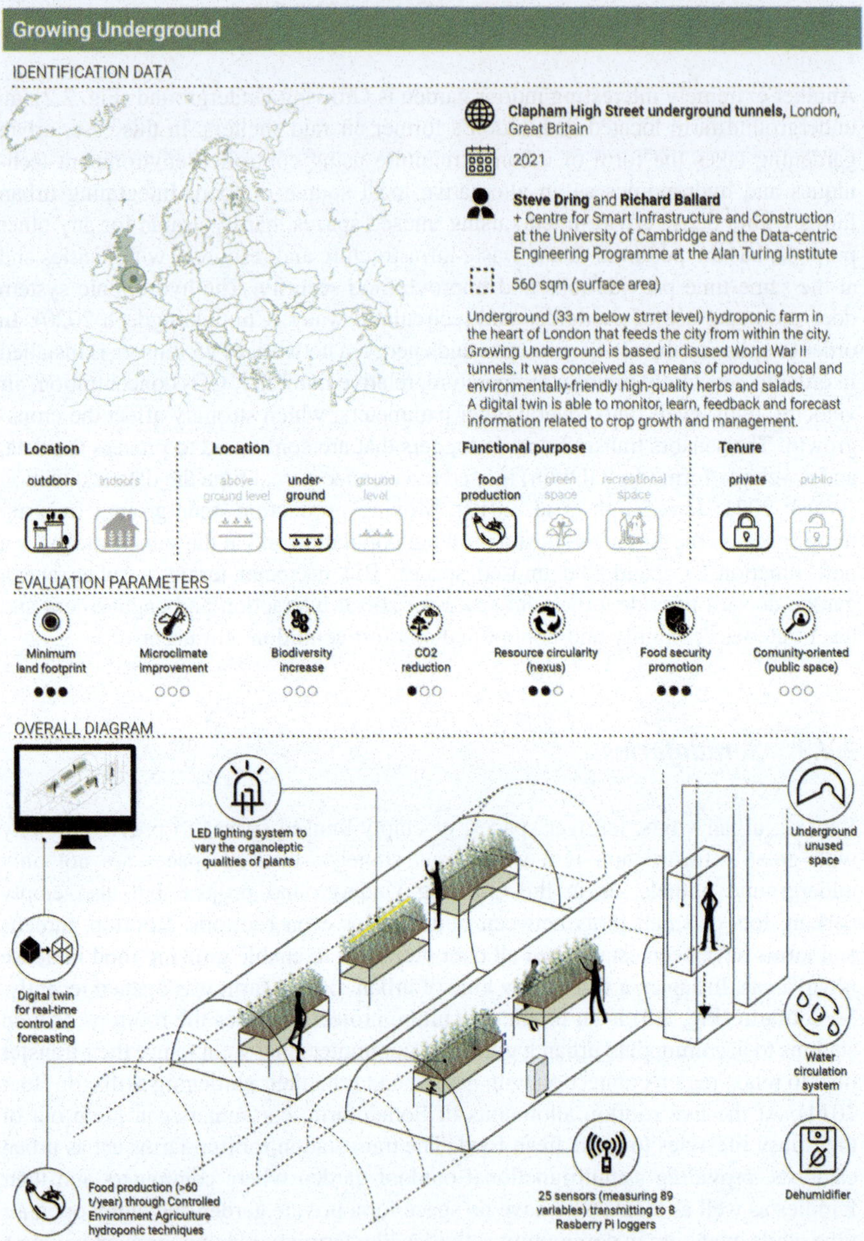

Fig. 2.2 Growing Underground in London's former air raid shelters in Clapham Highstreet underground tunnels: indoor hydroponic farming facility. (Source: Authors' elaboration)

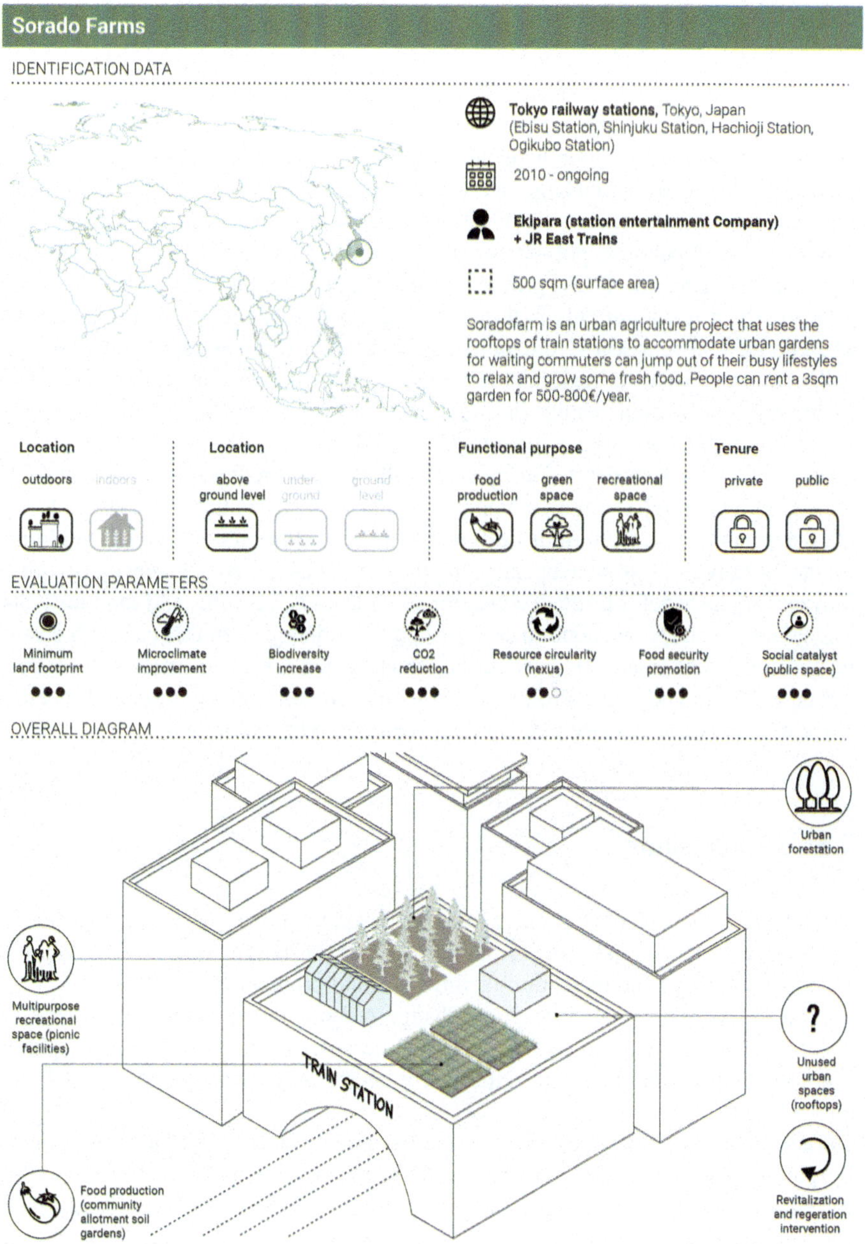

Fig. 2.3 Soradofarm on Tokyo's railway infrastructure: rooftop farming. (Source: Authors' elaboration)

2.4.4 Impact Farm

Since many cities around the world are pockmarked by empty, underused space, food deserts, and endemic underemployment, urban farms are gradually gaining ground as a potential solution to all such problems. An example of optimizing urban unused voids while enhancing the urban fabric in terms of social vitalization, microclimate improvement, and environmental reconversion is impact farm (Fig. 2.4), a greenhouse pop-up urban model that can easily squeeze into tight, urban spaces. The greenhouse provides a small-scale solution within 50 m^2 designed to be highly efficient not only in terms of food production but also in terms of water and energy consumption (Goldapple 2015). Irrigation is sourced from rainwater collection, and water is recirculated within a closed-loop system. The method of hydroponic cultivation results in significant savings in the use of freshwater at 70–85% compared to more conventional methods of production (UIA 2023). Moreover, the farm is designed for disassembly allowing for temporary leasing and site relocation, which represent a common challenge in metropolitan areas (UIA 2023). On the ground level, the farm is designed as a social space for community building. In the prototype built in Copenhagen, the farm has raised awareness of urban farming and its benefits through educational workshops, concerts, and food festivals. The goal is to create an economically sustainable urban garden that can both ensure resource-efficient local food production, create green jobs, and increase local economic activity (The Index Project 2023) while providing a neighborhood meeting place in a transitional space that is in between outdoors and indoors.

2.5 Conclusions

Overall, the concept of garden reveals to be quite broad encompassing indoor, outdoor, and transitional spaces; public and private tenure; and a wide range of functions ranging from recreation facilities to food production.

The indoor garden at Singapore Changi Airport and the rooftop gardens on Tokyo's train stations are conceived as public spaces within or on top of public mobility facilities, but in practice, they are intended only for privileged users. In the airport's case, the garden can be accessed by non-travelers only by paying an entrance ticket. In Tokyo, people can use the gardens and access the rooftops only by paying a yearly rent of about 500–800€. Because of its limited accessibility, linked to economic disparity, the "public" space loses its role as a social catalyst and its democratic nature.

The increasing demand for food, the lack of natural resources and arable land, and the urgency of limiting energy consumption have largely contributed to the evolution of the traditional garden into a food-producing garden. As shown by the case studies Growing Underground and Impact Farm, hydroponic farming is gaining popularity all over the world because of efficient resource management, land-use

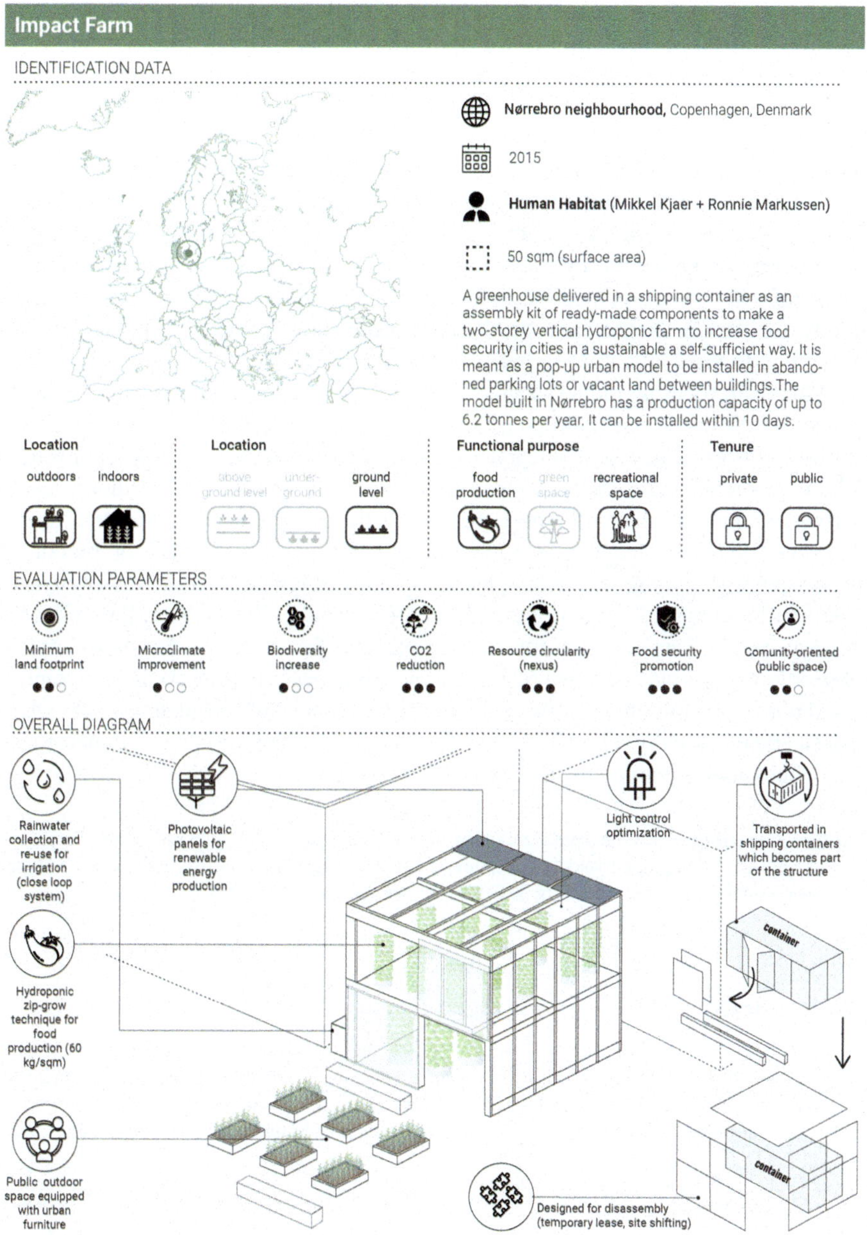

Fig. 2.4 Impact farm in Copenhagen: pop-up urban vertical garden for vacant and/or unused plots in cities. (Source: Authors' elaboration)

efficiency, planting environment cleanliness, fertilizer and resource saving, and quality food production. Soil-based agriculture is now facing various challenges such as urbanization, extreme weather events and natural disasters, and indiscriminate use of chemicals and pesticides, which is depleting the land fertility. However, as highlighted by the Soradofarms project, conventional cultivation requires less investment and technical know-how, making it more accessible for now. As a result, developing low-cost techniques that are easy to operate and maintain, require less labor, and lower overall setup and operational cost is crucial for successful implementation of commercial hydroponic technology. In light of these considerations, conventional cultivation methods and hydroponics and/or aeroponics techniques are likely to coexist in the coming decades, since they provide different benefits according to the context needs and function.

Moreover, focusing on the social role of gardens and green infrastructure, whether plants are grown using hydroponic techniques or soil is not important, since the purpose concerns the overall integrated benefits provided by the green space. Because of the unbearable temperature, humidity, lighting, and ventilation conditions, a subterranean garden will hardly be used as a gathering place or recreational space. However, thanks to controlled environment techniques, it is ideal for producing food. Although hydroponic techniques can be extremely beneficial for narrow spaces between buildings, along facades, and in other residual urban spaces, they may not always be the best solution for rooftop gardens for instance. As shown in the Soradofarm example, the green roof also serves as a means to improve indoor comfort conditions in the building, as the soil layer acts as a powerful insulating layer. These considerations highlight how different sites, in terms of morphological, dimensional, and microclimate conditions, require different solutions.

Author Contributions Conceptualization by Alessandra Battisti (A.B.); Introduction, Paragraph 1 and 2 by A.B., Paragraph 3 and Conclusions by Livia Calcagni (L.C.); Case study analyzed by L. C.; images and figures were made by L.C.; review was conducted by A.B.

References

Battisti A (2023) Urban regeneration between well-being, social determinants and sustainable development goals. In: Battisti IA et al (eds) Equity in health and health promotion in urban areas: multidisciplinary interventions at international and national level. Springer, Cham, pp 3–11

Beck U (2016) The metamorphosis of the world: how climate change is transforming our concept of the world. Wiley, London

Benedict MA, McMahon ET (2012) Green infrastructure: linking landscapes and communities. Island Press, Washington

Brown B, Aaron M (2001) The politics of nature. In: Smith J (ed) The rise of modern genomics, 3rd edn. Wiley, New York, pp 234–295

Chutiphongdech T (2020) The success behind the world's best airport the rise of the Changi. Asia Social Issues 15(1). https://doi.org/10.2139/ssrn.3715232

Data-Driven EnviroLab & NewClimate Institute (2020) Accelerating net zero: Exploring cities, regions, and companies' pledges to Decarbonise. https://datadrivenlab.org/wp-content/uploads/2020/09/Accelerating_Net_Zero_Report_Sept2020.pdf. Accessed 2 May 2023

de Boer J (2014, March 24) Pop up city, Japanese Commuters grow veggies on train station rooftops. https://popupcity.net/insights/japanese-commuters-grow-veggies-on-train-station-rooftops/. Accessed 1 Jul 2023

Doe J (2000) Title of supplementary material. http://www.privatehomepage.com. Accessed 22 Feb 2000

European Commission, Directorate-General for Research and Innovation (2015) Towards an EU research and innovation policy agenda for nature-based solutions & re-naturing cities: final report of the Horizon 2020 expert group on 'Nature-based solutions and re-naturing cities': (full version). Publications Office. https://data.europa.eu/doi/10.2777/479582. https://op.europa.eu/en/publication-detail/-/publication/fb117980-d5aa-46df-8edc-af367cddc202. Accessed 2 May 2023

Goldapple L (2015, November 26) Atlas of the future, The flatpack urban farm. http://www.springer.com/lncs, https://atlasofthefuture.org/project/impact-farm/. Accessed 1 Jul 2023

IPBES (2019) Global assessment report on biodiversity and ecosystem services of the Intergovernmental Science-Policy Platform on Biodiversity and Ecosystem Services. IPBES secretariat, Bonn, Germany, 1148 p. https://doi.org/10.5281/zenodo.3831673

IPCC (2023) Summary for policymakers. In: Climate change 2023: Synthesis report. A report of the intergovernmental panel on climate change. Contribution of working Groups I, II and III to the sixth assessment report of the intergovernmental panel on climate change. IPCC, Geneva

Jans-Singh M, Fidler P, Ward RM, Choudhary R (2019) Monitoring the performance of an underground hydroponic farm. In International Conference on Smart Infrastructure and Construction 2019 (ICSIC) Driving data-informed decision-making. ICE Publishing, pp 133–141

Moosmann L, Neier H, Mandl N, Radunsky K (2017) Implementing the Paris agreement–new challenges in view of the COP 23 climate change conference. Study for the ENVI Committee, European Parliament, Policy Department for Economic and Scientific Policy, Brussels

Munafò M (2021) Consumo di suolo, dinamiche territoriali e servizi ecosistemici. Edizione 2021. Report SNPA 22/21

Naumann S, Davis M, Kaphengst T, Pieterse M, Rayment M (2011) Design, implementation and cost elements of Green Infrastructure projects. Final report, European Commission, Brussels, p 138

Pitman SD, Daniels CB, Ely ME (2015) Green infrastructure as life support: urban nature and climate change. Trans R Soc S Aust 139(1):97–112

Raut A (2019, April 23) Design, Singapore's Jewel Changi Airport is home to the world's tallest indoor waterfall. https://www.architecturaldigest.in/content/singapore-changi-airport-worlds-tallest-indoor-waterfall-safdie-architects/

Sadin É (2018) L'intelligence artificielle. L'échappée, Paris

Safdie Architects (2023) Projects, Jewel Changi Airport. https://www.safdiearchitects.com/projects/jewel-changi-airport. Accessed 1 Jul 2023

Serres M (1990) Le contrat naturel. François Bourin, Paris

Taramanni E (2015, August 3) This big city, Soradofarms: gardening and community building for Tokyo Commuters. https://thisbigcity.net/soradofarms-gardening-and-community-building-for-tokyo-commuters/. Accessed 1 Jul 2023

The Index Project, Award Nominiees (2023) The impact farm. https://theindexproject.org/award/nominees/1176. Accessed 1 Jul 2023

UIA 2023 – World Congress of Architects (2023) Impact farm. https://uia2023cph.org/case-studies/impact-farm/. Accessed 1 Jul 2023

United Nations Convention to Combat Desertification (2022) The global land outlook, 2nd edn. UNCCD, Bonn. https://www.unccd.int/sites/default/files/2022-04/UNCCD_GLO2_low-res_2.pdf. Accessed 2 May 2023

Veerman C, Correia TP, Bastioli C, Biro B, Bouma J, Cienciala E, Wittkowski R (2020) Caring for soil is caring for life: ensure 75% of soils are healthy by 2030 for healthy food, people, nature and climate: interim report of the mission board for soil health and food: study

Walsh L (2021, March 29) Growing underground. https://www.cam.ac.uk/stories/growingunderground. Accessed 1 Jul 2023

Williamson KS (2003) Growing with green infrastructure. Heritage Conservancy, Doylestown, PA

World Economic Forum (2023) Nature and net zero. https://www.weforum.org/reports/nature-and-net-zero/. Accessed 2 May 2023

World Health Organization. Regional Office for Europe (2017) Urban green space interventions and health: a review of impacts and effectiveness. https://apps.who.int/iris/handle/10665/366036. Accessed 2 May 2023

WWF (2022) In: Almond REA, Grooten M, Juffe Bignoli D, Petersen T (eds) Living planet report 2022 – Costruire una società naturepositive. WWF, Gland

Chapter 3
Sustainable and Together: Between Ecology, Health and Governance

Romeo Di Pietro and Mattia Martin Azzella

Abstract This chapter examines the effects of biodiversity loss, which have dramatically highlighted how humanity is constantly dependent on ecosystem services even though the majority of the world's population doesn't even know what they are. This dependence is particularly evident in urban ecosystems, which are often characterized by a deficient and inefficient ecological network that is unable to provide sufficient services. We need to bring nature back to the cities to increase social well-being, to provide services to citizens and to ensure a greater resilience of cities, especially in this period in which the negative feedback of climate change is dramatically emerging. The statement "More nature in the cities" also means contributing to the conservation of biodiversity, which should become a primary objective in urban areas. Instead, very often it sounds like an empty slogan or an environmental policy issue restricted to protected natural areas only.

While the main goal is one, the solutions may be multiple and must necessarily be shared by involving city administrators, citizens and the scientific community. The skills of the scientific community will need to be different but complementary to propose a consistent pattern of multidisciplinary nature-based solutions. Accordingly, the specific contribution provided by each discipline should fit in with those provided by the other disciplines and be designed to maximize the effectiveness of the results. One of the most suitable solutions adopted by designers to counter the harmful effects of global warming is nature based and is known as "Green Infrastructure". The design of various types of green infrastructures is the basis of all sustainable development policies, both at the national and EU community level. If we are talking about green infrastructures, such as urban forests, rain gardens and green roofs, it is evident that they share a common element, namely the use of plants, species and/or plant communities. However, an essential rule should be followed, namely that the plant material used in green infrastructure projects would be better if it were of local origin and preferably consistent with the landscape and biogeographical context of the project sites. In addition to avoiding problems of genetic pollution

R. Di Pietro (✉) · M. M. Azzella
Department of Planning, Design, and Technology of Architecture, Sapienza University of Rome, Rome, Italy
e-mail: romeo.dipietro@uniroma1.it; mattia.azzella@uniroma1.it

F. Dal Falco (ed.), *Beyond the Garden*, Designing Environments,
https://doi.org/10.1007/978-3-031-50567-6_3

for the natural ecosystems surrounding the cities, this a priori choice would be ecologically sustainable and respectful of the identity of the places.

Plants are a great resource to heal the wounds that we ourselves are inflicting on our quality of life in cities, but they have their own identity card and, like all medicines that treat diseases, they have their own leaflet which cannot be ignored.

Keywords Biodiversity · Green infrastructures · Ecosystem services · Landscape

3.1 Once upon a Time, There Was Ecology

The last book published by Valerio Giacomini (1914–1981), who is among the greatest Italian and European scientific personalities, the father of Nature Conservation Management in Italy, a promoter of the Intergovernmental Program "Man and Biosphere" (MaB) and one of its most brilliant scientific animators and one of the first academics to make the scientific world aware of the importance of Natural Capital, is entitled: "Ecology because…" ("*Perché l'ecologia*" – Giacomini 1980). Thanks to his interdisciplinary culture, the author provides many fitting "ecological case studies" that he explained by associating scientific accuracy with philosophical and humanistic concepts. Although Giacomini provided answers on why ecological thought should prevail in governing our actions, especially when we are called upon to manage the environment, his explanations do not appear to have had much of an impact on subsequent generations. To be more responsive to current needs, it might have made more sense to entitle the book: "Why Ecology?" (*Perché l'ecologia?*) considering the behavior of the industrialized countries' governments (and public opinion itself), the competition to achieve the highest percentage increase in GDP, the denialism surrounding global warming forcefully supported by newspapers (in most cases owned by businessmen or multinationals), some doubts arise. When asking policymakers or people living in cities or isolated villages whether it is correct to follow the principles of ecology and give up gaining immediate wealth for individual countries through the intensive exploitation of natural resources so as to benefit humanity as a whole, the answer would certainly not be univocal. Or rather, it might be univocal and affirmative in words (which the wind blows away), whereas in facts based on concrete actions, it is clear that the ecological vision of the world remains a controversial topic, with the number of detractors equal to, if not higher than, that of the supporters. "Why ecology?" could be a perfect slogan for the COP28 summit, being held in Dubai in the United Arab Emirates (which is one of the leading producers of oil and greenhouse gas emitters), where organizers stated that there is no scientific evidence suggesting that fossil fuels need to be abandoned to achieve the objective of tackling the increase in global temperatures. This statement was made at a time in which the Global Tipping Points Report, a study carried out by the University of Exeter's Global Systems Institute, which coordinated 200 researchers affiliated with 90 research institutions in 26 countries, informs us that we are closer than we have ever been to satisfying 5 of the 25 established climate tipping points that are being monitored: the melting of glaciers in Greenland, in West Antarctica and of the permafrost covering large areas of the planet, the death of coral reefs in warm waters and the collapse of an ocean current in the North Atlantic. That is, the level beyond which a change, which is in this case due to global warming, becomes unstoppable and uncontrollable. Literally, a point of no return.

Ecology is an uncomfortable word that pinches the conscience and should, in some circumstances, either not be mentioned or be mentioned as an empty sounding board.

One of us (Di Pietro 2014), in the preface of an essay on the role of plant ecology in Architecture, started this way to introduce the "multitasking" and "controversial" meaning of ecology "*In the middle of the twentieth century, little was known about ecology. In the Seventies, either due to the activist campaigns on the alarming problem of pollution, or to address a real sense of guilt creeping into public opinion, the term "ecology" rose to the fore. In the nineties, ecology was the watchword for acquiring respectability, in all venues, institutional and otherwise, in all social and cultural gatherings, and in the countless political debates. It was enough to pronounce the term "ecological" in a discussion or in an electoral meeting to automatically pass onto the side of the "goodies". Today we hardly hear about "ecology" anymore because it is no longer a trend, having been replaced by terms that are at least apparently less binding but more engaging and captivating, such as "biodiversity conservation", "green economy", "environmental quality"*.

On the other hand, the history of ecological sciences is dramatically characterized by the awareness of lost things. Alexander Von Humboldt, observing the South American plains at the end of the eighteenth century, was the first to highlight how man was disrupting the *Naturgemälde* (Wulf 2015). Von Humboldt suggested that "there is a chain of affinity binding together all nature" and that the intimate connections that characterized natural environments were at risk. This happened several years before Haeckel invented the word "ecology" in 1866 and founded the science that investigates the relationships between the physical environment and living matter together with the connections (spatial, functional, temporal) that occur within the latter. These are "the connections of nature", the functioning of our οίκος (Levit and Hossfeld 2019). An οίκος (home) that, at the end of the nineteenth century, had its foundations undermined and was going up in flames. Thirty years after the definition of ecology, in 1896, Arrhenius laid the foundations of modern climatology (Rodhe et al. 1997), a science born to outline the impact of man on nature. Arrhenius was the first to prove that carbon dioxide introduced into the atmosphere by industrialization would accumulate over time, causing a rise in temperatures. In 1962, Rachel Carson proved the effect of pesticides on the biodiversity of birds (Carson 1962). In 1972, the "Club of Rome" highlighted the limits and risks in the growth of the world population (Meadows et al. 1972). Apparently, everything was already clear, everything was written and demonstrated many years before the birth and spread of the word "biodiversity", in 1986. This word was coined to define the diversity of nature and to remind humanity that such diversity was precisely what we were rapidly losing. One year later, the concept of "sustainable development" was defined by the Brundtland Commission. The definition was born to call for a change in economic development to meet the needs of future generations without leading to a depletion of environmental resources. The history of the last 35 years, unfortunately, has shown that inserting "sustainable" in front of the word "development" has not brought the desired results. The development was unsustainable in 1987 as it still is today.

In 1992, the "United Nations Conference on Environment and Development" was held in Rio de Janeiro and, considering the impact of the event, there were all the conditions for it to represent a turning point in the establishment of global awareness on the value of biodiversity and on the implementation of effective policies for its protection. Europe tackled the problem of biodiversity loss by ratifying the Habitat Directive 92/43/CEE. In the package of measures and rules constituting the core of the Habitat Directive, for the first time, attention was shifted to the management of ecosystems rather than to their conservation tout-court. In fact, the Habitat Directive put forward the innovative concept that some anthropic activities could even serve as bulwarks to face the loss of biodiversity, especially in the so-called semi-natural environments. Unfortunately, despite these interesting and innovative environmental policies, the loss of biodiversity did not slow down.

In 1997, the concept of Ecosystem Services was born. Once again, a branch of ecology arose to define what we were losing. In their scientific work that represented a new milestone in nature conservation, Costanza et al. (1997) showed that the loss of biodiversity, in terms of species and ecosystems, would lead to the loss of numerous functions that the ecosystems themselves performed and that could be easily quantified as services for humanity. Since then, thousands of scientific papers have been published on this topic (Turner et al. 2007), and ecosystem services have become the backbone of European strategies for the safeguard of biodiversity (Maes et al. 2020). The introduction of the concept of ecosystem service which evaluated not only qualitatively but also quantitatively the useful contribution of plant communities at different scales (from the oak forest stands to be planted in the surroundings of great cities to the micro-garrigues rich in *Sedum* sp.pl. and therophytes used for green roofs) represented a further leap towards a more functional and, in some ways, "productive" view of the environment. It became clear that a vigorous biodiversity was required to counteract climate change, to feed a growing population and to maintain a sustainable socio-economic fabric which at the same time ensures a high quality of life. Taking this for granted, the first action to be taken was (and still is) that of reversing the current, and unfortunately excessively fast, trend in biodiversity loss.

The MAES (Mapping and Assessment of Ecosystems and their Services) project, completed within the framework of the European Strategy for Biodiversity for 2020, has defined and evaluated the state of European ecosystems (Maes et al. 2020). Urban ecosystems are growing. Although not completely devoid of biodiversity, urban environments (especially large cities) have seen a drastic decrease in green spaces. Thus, the enhancement of nature in cities has become a priority. One of the strategies hypothesized by ecologists, who promptly suggested it to city administrations, was that of reconnecting urban spaces to the surrounding natural areas in order to reduce habitat fragmentation, the latter phenomenon being one of the leading causes of biodiversity loss at both the global and local scale. Unfortunately

(...for whom?), this goal could only be effectively achieved by moving in a diametrically opposite direction to how productive society is currently moving, namely by preserving the natural ecosystems that survived the expansion of artificial surfaces, restoring the most severely compromised natural spaces and planning different types of green infrastructures at different scales. However, while the objective is clear and the strategies may share similar bases in different countries, the methodologies (technical or scientific) to be used cannot be generalized. In other words, while there is one globally recognized problem, the solutions may be numerous but inevitably need to be found locally. Let's take the example of urban forests, the most popular solution proposed to tackle the heat island effect in cities. Every urban forest is a living and dynamic system bound by the rules of the evolution path and by the limits imposed by biogeographic maps. Therefore, if the suggested solution is "planting trees", then we must consider that there are countless connections that have developed in natural ecosystems and that have been perfected (phenotypically and genotypically) over hundreds of thousands of years. The genetic variability of an individual (or a population) is a resource that allows the resilience of species and ecosystems and should be primarily considered when plants are going to be used in green infrastructure. To sum up, each place will have its most suitable tree choice (or association of trees in the form of a wood stand) which will be different from the most suitable tree choices in other places, these "other places" presumably being characterized by different ecological, climatic and biogeographical conditions. In short, "the right tree for the right place".

3.2 The Wealth of Biodiversity

Biodiversity is a multidimensional concept. The simplest way of understanding it, which will be clear to everyone, is to count the number of species[1]: a biodiverse ecosystem tends to host many species. For instance, a patch of 10 km^2 in an intertropical rainforest can contain more than 2000 plant species. Not all the ecosystems in the whole Biosphere are characterized by such high plant biodiversity. However, the specific richness of an ecosystem is not the only parameter to consider when assessing its importance. In fact, there are many poor-in-species ecosystems (e.g., those developed in environmental conditions in which one or more limiting factors act, such as the salty steppes of coastal areas – Fig. 3.1), among the rarest and most vulnerable ecosystems in the world, in which a microclimatic change or a slight increase in anthropic pressure could lead to their local extinction.

[1]There are many methods for counting species and therefore assessing the biodiversity of an ecosystem. Many indices have been developed and even the simple "number" of species is a complicated and multidimensional concept. The scientific literature on the subject is vast, and the works we cite (Moreno et al. 2018; Tuomisto 2010) are but a few examples.

Fig. 3.1 A submersed meadow of Chara aculeolata. Stonewort's meadows are an example of poor-in-species ecosystems, which are threatened worldwide

The spatial arrangements of a high number of poor-in-species communities related to the sharp change in micro-climatic and micro-topographic conditions are the most typical form of ɣ biodiversity (number of ecosystems per unit area). In the field referred to as "Vegetation Science", ɣ biodiversity assumes greater value when it is composed of a highly diverse pattern of potential plant communities (high potential heterogeneity), which means that the plant landscape is composed of a high number of "climax" (potential natural vegetation) communities. Very often, especially in environments that have interacted significantly with the presence of humans, an apparently high "actual" heterogeneity of the landscape does not correspond to an equally high "potential" heterogeneity. From a bird's-eye view, the Po river valley, in northern Italy, currently looks like an intricate puzzle of intensively cultivated fields with various crops for dozens of square kilometers. In terms of potential heterogeneity, however, it refers to just a single type of potential natural vegetation, i.e., the alluvial forest of *Quercus robur, Q. petraea, Carpinus betulus* and *Ulmus minor* (Blasi 2010). By contrast, by following a linear spatial transect of only 100 m crossing a dunal environment from the shoreline to the back of the dune, a spatial succession composed of different "permanent" plant communities can be observed, each of which is dominated by a single species (*Cakile maritima, Elymus farctus, Ammophyla australis, Crucianella maritima, Juniperus macrocarpa,* etc.. . .) and each of which acts as a single type of potential natural vegetation (Géhu et al. 1984).

Examples of the importance of poor-in-species ecosystems can obviously also involve the fauna and be read from the point of view of the entire food web. The ecosystem that allows salmon reproduction is very poor in species: a stream with some bare rocks. However, salmon belong to a keystone species that is needed by many other animals and ecosystems (Reimchen 2000).[2] Without Pacific salmon, wolves, bears and scavenging birds do not have enough food (Willson and Halupka 1995). The local extinction of salmon affects the forest ecosystems close to the rivers where the salmon spawn, which means they eventually collapse. Thus, the ecosystem diversity that allows salmon to survive is crucial. Accordingly, to protect the upper stream, near the source, where salmon mate, all the ecosystems of the river are important, even if they appear to be poor in species. Many other examples exist that highlight the pivotal role of keystone species and ecosystems, such as the mangrove forests that protect the tropical seacoasts (Kelleway et al. 2017) or the previously mentioned dune vegetation communities dominated by *Ammophila australis*, which are so important to the functionality of dune ecosystems (Feagin et al. 2015).

The variability between different species becomes clear when we analyze animals, whereas the genetic variability within species is immediately evident when we observe the multiform phenotypic expressions manifested by the species (*Homo sapiens*). Instead, phenotypic and genotypic variability (especially infrageneric variability) tends to be somewhat ignored when it comes to plants. Our yardstick, which is calibrated to identify only macroscopic differences, might lead us to conclude that all the birches or deciduous oaks of the world must belong to the same species. Taking birch as an example, we might think that the disappearance of birch from the slopes of Mount Etna in Sicily would not be a great loss since birch (*Betula pendula* L.) is probably the most widespread deciduous tree species in central and northern Europe (Euro+Med 2006). It would actually be a painful loss regardless (at least for the two authors of this chapter), but it would be an even more painful loss if we consider that the Etna birch populations do not belong to the species *B. pendula* found throughout Eurasia, but to a taxonomically separate species, *Betula etnensis* Rafin, whose occurrence in the world is exclusive to Mount Etna. It is for this reason that observing *Betula etnensis* in its natural ecosystem immediately and unequivocally places us on the geographical map and allows the Etna landscape to be immediately identified and circumscribed with respect to the landscape of the rest of Sicily and of any other place in the Italian peninsula. In addition to the multiform aspects of anthropic interactions between human activities and natural ecosystems, there are millions of other cases in which biotic interactions between living organisms lead to significant (but sometimes invisible) modifications in the environment. These may appear less spectacular than those that include man among the modifying agents, and yet they are extremely significant for both ecosystems and the landscape. We should perhaps finally realize that, whether we're talking about *Homo sapiens* and *Betula etnensis* or about *Ailuropoda melanoleuca* (giant panda), *Canis lupus* (wolf), *Lumbricus terrestris*

---

[2] salmon: a keystone species (https://pacificwild.org/salmon-a-keystone-species).

(earthworm) or *Latrodectus mactans* (black widow), they all have the same value in taxonomic terms (each of them counts for one single species). In broad terms, the extinction of any of the aforementioned species would have the same negative impact on the computation of the overall alpha-biodiversity of our planet and should cause us the same displeasure, regardless of their external appearance. We are aware that the above comparison may sound quite trivial, especially if we look at it from the perspective of mankind. However, the combination of the "equal specific dignity" with what was said a few lines above about the importance of the Etna birch allows us to introduce the concept of biogeographic identity of plant species and communities, which is in turn strictly related to those of landscape identity. Biodiversity is important because the higher the number of species, the more functions ecosystems perform (Tilman et al. 2014). For example, a meadow with a high number of species has a high aboveground biomass (Spehn et al. 2000). That means it produces taller hay for livestock than a meadow characterized by a lower number of species, capturing and storing more carbon dioxide and allowing the survival of more bees. Thus, the higher the biodiversity, the greater the number of ecosystem services available. Nevertheless, there are normally very few species in our cities, in public parks and gardens as well as in private backyards. Biodiversity is normally made up of a much smaller number of species than the environment could naturally host, and it very often contains introduced species (alien) or obviously artificial associations of a few species. Biodiverse meadows are decreasing and butterflies are disappearing, as are bees. In order to change the approach in the design of green areas and go beyond the garden, knowledge of the "concept of biodiversity" is required, as is an awareness of what biodiversity is currently under threat. Biodiversity is rapidly collapsing, and it is exclusively because of us. This is the biggest global problem we are facing, a problem that is even more alarming than climate change, to which it is in any case closely connected. Indeed, lots of the scientific evidence and figures on the biodiversity loss highlighted by IPBES[3] are more bewildering than the alarming numbers related to the negative effects of climate change. The number of insects, the most species-rich group of living beings, is collapsing worldwide (Sánchez-Bayo and Wyckhuys 2019). In Europe, there has been a 76% loss in insect biomass (Van Klink et al. 2020). Even if insects may disgust us a little (and once again. . .the authors of this paper do not think so), "we will miss them" (Kolbert 2020). It is well known that insects and plants have co-evolved, and many Angiosperms (plants producing flowers) need pollinating insects. Thus, the 50% reduction in butterflies (pollinating insects) will generate a cascade effect on the efficiency of pollination on flowering plants, the results of which can easily be deduced in terms of seed and fruit production and as intensity and abundance of the subsequent phenological cycles. This is due to the fact that flowering plants are the reason why biodiversity is so abundant on our planet (Benton et al. 2022).

---

[3] Intergovernmental Science-Policy Platform on Biodiversity and Ecosystem Services: https://www.ipbes.net/

Recent studies on the Earth biomass have outlined something shocking. The extinction rate of mammals, birds and fish has skyrocketed exponentially over the last 100 years, and the mass of all mammals and wild birds, from mice to whales, from hummingbirds to condors, combined currently accounts for only 0.38% of the planet's total biomass, while humans account for 2.5% and farmed animals for 4%. On Earth there are now more chickens on farms than birds in the wild, in a ratio of about 100 to 1. We have disrupted biodiversity, and this is a huge problem because (and we apologize for underlining the following concept so frequently) biodiversity is essential to our life.

3.3 Ecosystem Services in a Changing World

Biodiversity provides us with ecosystem services, which are the benefits that human beings get from nature. It is important to note the slight difference between ecological functions and ecosystem services. We will try to explain it with an example. Through a series of complex anabolic chemical reactions known as photosynthesis, plants produce sugar from carbon dioxide and water. The byproduct of this reaction is oxygen. All the oxygen that we and all other animals breathe in every second of our lives is made available to us free of charge. Thanks to plants, the level of carbon dioxide in the atmosphere of planet Earth is lower than that in the atmosphere of Venus, a planet that is similar to Earth insofar as it is of a similar size and has a similar position to ours in the solar system. The greenhouse effect present on Earth, which causes us so much concern, has nothing to do with the extreme greenhouse gas effect on Venus, which is incompatible with life itself. Obviously, plants perform photosynthesis not to please us but to fulfil a function designed to produce glucose for the plants' own nutriment. However, if we look at the amount of carbon dioxide that plants remove from the atmosphere thanks to the activity of an urban forest and we compare and economically quantify it with a job (duty) aimed at reducing the rising temperature, we are assessing an ecosystem service. Thus, once the function (plants produce fruit to spread seeds) is viewed as a service for us as human beings (we collect, sell and eat fruits), we move from the concept of ecosystem functions to that of ecosystem services (ES). Having ascertained that ecosystem services exist and are given to us on a plate, let us ask ourselves which places on Earth need these services most.

The MAES project mapped and assessed, between 2010 and 2020, the European ecosystems and the services they provide within the framework of the European Strategy for Biodiversity. Among the various European ecosystems considered, MAES identified and more thoroughly analyzed one in particular, which could only marginally be classified as a natural origin ecosystem: the urban ecosystem. In this case, we should take a small step back and move from the scientific concept of "ecosystem" to the more popular one of "environment". The environment, using basic terminology, is "everything that is around or that surrounds something". Normally our ego leads us to formulate the concept of "environment" according to an anthropocentric vision: the object of the environment is the human being. However, it is not easy to establish what habitat provides the ideal conditions for the existence of humans, or, using a slightly more scientific terminology, the habitat in which both the autoecological and the synecological optimum coexist for the species *Homo sapiens*. It is very likely precisely that mass of inert materials that are artificially modeled, sometimes with skill and intelligence, other times somewhat illogically (…except the logic linked to business and short-term profitability), and very often assembled with no ecological and landscaping foresight. In short, we are just saying that the beautiful place that goes by the name of "urban environment" is probably the one with which human beings are most frequently associated. In this unnatural inert material–human being combination, the maximum population density and the maximum phenotypic and genotypic diversity for our species (*Homo sapiens*) are found at present. The curious thing is that cities, which are arid and life-impermeable, at times unbreathable, environments, are for many of us indispensable if we consider that at least 50% of the world's population lives in cities and about 10% of it lives in the largest metropolises. Despite accounting for a negligible percentage of the Earth's surface, the urban environment is among the most familiar to man, certainly among the most easily identifiable and at the same time among the most investigated in scientific terms. This environment is actually expected to become the only one possible for a significant proportion of young future generations, given that migration from the countryside to the city is estimated at around 60 million people a year, and that some recent projections predict that by 2050, over 80% of the world's population will live permanently in large cities.

Despite the different origins and the evident different physiognomy and outer color between the urban ecosystems and the more natural ones, we can find similar ecological dynamics in both cities and natural environments, albeit with some exceptions.

Urban ecosystems, like the natural ones, are characterized by a well identifiable core area, which in the case at issue is the center of the city. Associated with the core area there is a commuting zone, which in most cases is far larger than the core area, where most of the population that reaches the core area every day to work lives. Forming an external belt, concentric to both the core area and the commuting zone, and with an extremely variable radius, there is the so-called functional area, which is the portion of the territory providing the resources on which both the core area and the commuting zone depend. It is therefore clear that, in ecosystem terms, both the city core area and the commuting zone will act as the main users of ecosystem services, whereas the functional zone will represent their producing source. In recent years, there has been a lot of focus on how to improve the quality of life in cities. In fact, the design and implementation of green infrastructure (such as urban forests, green roofs, rain gardens) is almost always dedicated to core areas or commuting zones. Although there is absolutely nothing wrong with that, the well-being of natural areas outside cities should be managed and safeguarded with the same level of attention. Indeed, the proper functioning of cities, or their ability to resist and to develop resilience to global changes, will depend precisely on the resources created and put into circulation by functional areas. Let's take the example of the resource "water". The water a city (core area) needs comes from sources (functional zones) often located a long distance from the city core. Water consumption is growing worldwide, with European cities being no exception. Every year cities need an ever-increasing quantity of water, both to carry out production activities and for daily domestic needs. Water supplies are in danger, not only because of overexploitation but also owing to climate change. Accordingly, the water supplies of a city can be ensured only by protecting the well-being of the natural environments that make up the functional zones (e.g., forests, shrubland, grasslands, mires, marshes, rivers, etc.), thereby allowing the water tables to recharge with water. Planning must therefore go beyond the boundaries of the core of the city and ensure the ecological connection between the city and the surrounding natural ecosystems by protecting them.

3.4 Global Problems, Local Solutions

The urban ecosystem is the most widespread in Europe, as highlighted by the EU ecosystem assessment (Maes et al. 2021), and it is the only ecosystem that is constantly growing. Although there is some good news, such as the improvement in bathing water quality and the reduction in emissions of air pollutants, many other problems persist: an uncontrolled invasion of alien species; +11.3% share of dispersed settlements in peri-urban areas; −4.4% vegetation loss inside urban green space of core cities; and − 6.4% vegetation loss inside urban green spaces in the commuting zone. In a decade (2010–2020) in which the population and the governments of the countries started to become seriously aware of the importance of biodiversity for ecosystem services and the need to restore ecosystem services, we witnessed a continual loss of nature in our cities, in which existing green infrastructures were dismantled as opposed to new ones being created. In fact, we continue to artificialize considerable portions of European territory, threatening biodiversity, and therefore threatening ourselves and our survival.

Urbanized areas share common problems, i.e., poor air quality, high levels of noise pollution, limited capacity to tolerate flooding events (and increased risk of flooding due to the high share of impermeable surfaces), to cope with the urban heat island effect and, finally, an excessively modest contact with "natural" environments and green spaces. We need to increase our effort to defend biodiversity, and one way to start this defense action is by "bringing nature back into our lives".[4] Since 72% of the European population currently lives in urban areas, "bringing nature back into our lives" means bringing it back into our cities.

[4]The motto of EU 2030 Biodiversity strategy. Document 52020DC0380. Communication from the commission to the European parliament, the council, the European economic and social committee and the committee of the regions EU Biodiversity Strategy for 2030 Bringing nature back into our lives. COM/2020/380 final. link

The Natura 2000 network is the backbone of the European green infrastructure[5] and the biggest network of protected areas in the world. The aim of the network is to ensure the long-term survival of Europe's most valuable and threatened species and habitats, listed under both the Birds Directive and the Habitats Directive. Reconnecting cities to the Natura 2000 network is one of the main objectives of the European community and the task set for the EU governments in the next 20 years. At present, the Natura 2000 network protects 18% of the European territory, but the goal is to reach 30% by 2030 and the challenge is precisely to achieve this increment by strengthening this network in our cities and improving urban naturalness. How can that be done? The most immediate solution would to recover marginal areas and restore natural ecosystems where they were once present. However, there are also other hybrid solutions, such as green roofs and city parks devoted to biodiversity and not just to the enjoyment of the citizen, which will, even if they do not enhance the Natura 2000 network, certainly contribute to a greater efficiency of the local or regional ecological network.

3.5 Flora, Vegetation and Landscape Identity

To conclude this dissertation, let's take a small step back and return to a concept we have already mentioned in the previous pages. The problems are common, just as the solutions are very often common, but the application of these solutions cannot follow the same pattern because each of the places in which these solutions need to be applied has its own identity.

In order to address climate change and to halt biodiversity loss, a huge global effort will be needed over the next 20 years. Unfortunately, we will not be able to limit ourselves to conserving the most beautiful and precious natural communities that are still present, but we will be called upon to restore entire ecosystems. For example, ecosystems that ensure high resilience and protection from rising water levels will need to be restored on the coasts. Freshwater ecosystems, the most at risk globally, will have to be thoroughly restored because floods and droughts are two sides of the same coin (climate change) that make vast territories of entire countries fragile. Recreating the pattern of permanent micro-communities that characterizes coastal dunes, restoring the ecological balance in a river and setting up urban forests are three examples that require joint planning by a range of experts in different professions. In Italy, the native vascular flora consists of 8237 species. The total number of plant species (native and alien species) consists of 9897 taxa, belonging to 1547 genera and 198 families (Bartolucci et al. 2018). The past paleoclimatic and paleogeographic changes and the current very diversified bioclimatic, lithological and orographic pattern of the Italian peninsula allow the development of hundreds of different habitats and ecosystems. Each ecosystem has its own features. Thus, when

[5] Link to the European website about the green infrastructures.

we plan to create a new green infrastructure in a city or we decide to restore a green infrastructure that has lost its original degree of biodiversity, we must carefully plan our interventions. In the design of urban forests, the same association of tree species, scrub species and herb species cannot be used in Milan, Rome and Palermo. The urban forest (which is a common solution to a common problem) in Rome will have its own specific composition which will correspond to what is already present in the natural forest ecosystems present in the city center of Rome or in its countryside.

Therefore, the synergy between botanists and architects is fundamental and essential to combine the ecological characteristics of a plant community with the concept of landscape identity and the identification of places. In fact, vegetation is not distributed randomly, and different species that have similar ecological needs are associated according to ecological rules (Braun-Blanquet 1964). Species and communities derive from an evolutionary path lasting millions of years, a path characterized by biogeographical and climatic barriers that today are reflected in the current geographical distribution of the species. It is precisely this background that provides the landscape with an identity card, a millenary identity that could be erased by reckless planning choices.

It is very often the lack of attention paid to the microscopic and sometimes invisible interrelationships that are established between various forms of living matter within a territory that results in the need to address difficult macroscopic environmental problems. "An action here and now produces an effect there and then" reads the first law of Landscape Ecology (Forman and Godron 1986). Nothing is truer. This is the case of the thoughtless use of alien species in environmental design projects of urban parks, or in a simple botanical arrangement of a public (or private) garden. Indeed, the planting of species, greening of flowerbeds and reforestation actions in urban or suburban areas are often carried out without thinking in the slightest of the consequences that the introduction of an alien species could bring to the surrounding natural areas. Only rarely is the degree of invasiveness of an alien species taken into account in the design stages in setting up parks and gardens and very often, dazzled by an assortment of exotic scents and colors, one is unaware of how harmful the negative implications of a "local" negligence of this kind might be on a larger scale. For this reason, a greater degree of integration between landscape architects, taxonomists, phytosociologists, ecologists and other scientific figures would be desirable both in environmental design and in planning. Unfortunately, this virtuous mixture of scientific expertise is often absent, which means that from the very first stages of a landscape plan, in which the use of plants should assume a leading role, some fundamental aspects of the "instructions for use" are missing, and nefarious biotic interactions between species may occur. It is no coincidence that the native *Calamagrostis arenaria* and *Elymus farctus* communities in the dune ecosystem are rapidly and aggressively being replaced in the Mediterranean coastal landscape by the colorful and luxuriant communities of the succulent South African *Carpobrotus acinaciformis* or *C. edulis* as a consequence of the latter's inconsiderate use in cottages, villas and beach resorts. Nor we can ignore the regression of *Pistacia lentiscus, Myrtus communis, Phillyrea latifolia, Arbutus unedo, Calicotome villosa* and many other native taxa in the native Mediterranean

Fig. 3.2 Colonization of the back-dune of the coast of Taranto (Puglia region southeastern Italy) by the hottentot-fig (*Carpobrotus acinaciformis*), an alien South African species that is about to surround a nucleus of large-fruited juniper (*Juniperus macrocarpa*)

maquis biome in various European Mediterranean countries owing to the invasion of some American prickly pear cactuses (e.g., *Opuntia Ficus-indica*) (Fig. 3.2).

The surviving lowland oak woods with *Quercus robur* and *Carpinus betulus*, reduced to the margins of the valley floors of the peri-urban environment of the city of Rome, offer only weak resistance to the attack of the black locust tree (*Robinia pseudoacacia*) or paradise tree (*Ailanthus altissima*) and of all those competitive non-native species with a very high invasive capacity that find their best allies for complete success precisely in the disturbance and degradation. The overall picture is dark. We are already very late in reaching a full awareness of the danger, and unfortunately, the administrations that govern us do not appear to be adopting any preventive measures to tackle the problem.

This progressive invasion of alien species doesn't resemble a coup, nor is it loud. It is a silent advancing tide. The only perceivable echo is that of the identifying characteristics of the places that inexorably evaporate into thin air.

The number of non-native species is growing drastically in the sites that are affected most by man, such as urban, industrial and agricultural areas, because such species are favored by the high degree of anthropic disturbance and by the greater contribution of propagules (Di Castri et al. 1990; Pyšek 1998). Biological invasions, i.e., the uncontrolled spread of species transported by man beyond their natural dispersion limits, are considered one of the main components of global changes and cause enormous damage, including those inflicted upon crops due to weed species, those inflicted upon people's health by pathogens, allergenic species, parasites, etc., and those inflicted upon ecosystem services due to changes in their functionality. According to the United Nations Development Program (UNDP), the invasions of alien species are the second biggest cause of the loss of biodiversity in the world after the direct destruction of habitats.

Plant species have a name and a surname which is linked to hundreds of thousands of years of evolution, of migrations across lands now covered by seas, of settling in glacial shelters, of separations due to the emergence of mountain ranges and much more. For this reason, the presence of a species in its area of origin has an infinite value, and being able to identify it in its natural environment comforts us and immediately positions us in space and time.

Flora and vegetation as an oasis of memory... precious treasures.

3.6 Conclusive Remarks

Designing a green infrastructure, even if it is merely a flower bed, a green roof or a rain garden, must start from the knowledge of local biodiversity. To get an idea, we can refer to the technical annexes of the recent urban reforestation call promoted by the Italian Minister for Ecological Transition. The Urban Reforestation program aims to plant 6,600,000 trees and shrubs according to the principle of using "the right tree in the right place". The call provides information on the characteristics of the vegetation and therefore of the local ecosystems. The experts called upon to carry out reforestation are expected to conduct preliminary studies regarding the potential natural vegetation of the sites and the different stages of the Vegetation Series. Indeed, the different metropolitan areas have different vegetation series. Therefore, for each city (and in different areas within the same city), there are different "lists" of useful trees (Fig. 4). This advanced concept, which considers the neutral model adopted by nature (potential vegetation) as the most ecologically and economically advantageous example from which to take inspiration, is still partially unclear, but we must keep it in mind. It is not a habit; it is not an unnecessary complication. It is the starting point for successful green infrastructure planning (Fig. 3.3).

On the other hand, we have long since reached a point of no return with the environmental issue. It is a real social emergency for modern society, and dealing with this emergency has become a duty and an obligation for every single citizen. As decreed by the European Community to all national governments, in order to plan the sustainable management of resources on a global scale, it is first necessary to plan and implement multidisciplinary interventions on actions on a local scale. It is therefore desirable that future environmental restoration and recovery interventions, as well as all landscape plans and master plans, that significantly affect landscape

Fig. 3.3 Wheat field mixed with poppies (*Papaver rhoeas*) on the outskirts of Rome with a specimen of cork oak (*Quercus suber*) in the center of the photo, a solitary bulwark bearing witness to the "neutral model", in this case, the *Quercus suber* evergreen thermophilous forest, i.e., the potential vegetation of that area if there hadn't been any crops

management, be the result of the interaction between different disciplines (ecological, humanistic, technical-scientific, architectural). Each of these skills will be the bearer of its own interpretation, which should remain complementary in its analysis but needs to provide results that can be exported and prove useful to other disciplines while gathering innovative ideas and application possibilities from the latter. The systemic concept of "landscape" itself frames it, albeit not intuitively on an exclusively aesthetic-perceptive basis but as a "set of interacting elements" within which several disciplinary components combine to characterize its profile.

Environmental design in the field of architecture arises precisely from the awareness of the multidimensional "surrounding reality", from the need to relate architectural aesthetics to the environmental matrix, so as to avoid reducing it to a mere visual experience.

Author Contributions Both authors designed the study and wrote the first draft of the paper.

References

Bartolucci F, Peruzzi L, Galasso G, Albano A, Alessandrini ANMG, Ardenghi NMG et al (2018) An updated checklist of the vascular flora native to Italy. Plant Biosyst 152(2):179–303. https://doi.org/10.1080/11263504.2017.1419996

Benton MJ, Wilf P, Sauquet H (2022) The angiosperm terrestrial revolution and the origins of modern biodiversity. New Phytol 233(5):2017–2035. https://doi.org/10.1111/nph.17822

Blasi C (a cura di) (2010). La Vegetazione d'Italia. Palombi & Partner S.r.l. Roma

Braun-Blanquet J (1964) Pflanzensoziologie. Grundzüge der Vegetation-skunde. Springer, Wien. https://doi.org/10.1007/978-3-7091-8110-2

Carson R (1962) Silent spring. Houghton Mifflin Harcourt, New York

Costanza R, d'Arge R, De Groot R, Farber S, Grasso M, Hannon B et al (1997) The value of the world's ecosystem services and natural capital. Nature 387(6630):253–260

Di Castri F, Hansen AJ, Debussche M (eds) (1990) Biological invasions in Europe and the Mediterranean Basin. Kluwer Academic Publisher, Dordrecht

Di Pietro R (2014) Plant ecology in environmental design and planning: lecture 4 (Italian). Lectures of the Department of Design, planning and architecture technology Rdesignpress (www.rdesignpress.it). 74–95

Euro+Med (2006) Euro+Med PlantBase - the information resource for Euro-Mediterranean plant diversity. Published on the Internet http://ww2.bgbm.org/EuroPlusMed/. Accessed on August 2023.

Feagin RA, Figlus J, Zinnert JC, Sigren J, Martínez ML, Silva R et al (2015) Going with the flow or against the grain? The promise of vegetation for protecting beaches, dunes, and barrier islands from erosion. Front Ecol Environ 13(4):203–210. https://doi.org/10.1890/140218

Forman RTT, Godron M (1986) Landscape ecology. Wiley, New York

Géhu JM, Costa M, Scoppola A, Biondi E, Marchiori S, Peris JB (1984) Essai synsystématique et synchorologique sur les végétations littorales italiennes dans un but conservatoire. *Documents phytosociologiques* 8:393–474

Giacomini V (1980) Perché l'ecologia. Edizioni La Scuola, 192 pp

Kelleway JJ, Cavanaugh K, Rogers K, Feller IC, Ens E, Doughty C, Saintilan N (2017) Review of the ecosystem service implications of mangrove encroachment into salt marshes. Glob Chang Biol 23(10):3967–3983. https://doi.org/10.1111/gcb.13727

Kolbert E (2020) Where have all the insects gone? National Geographic. 23 April 2020. https://www.nationalgeographic.com/magazine/article/where-have-all-the-insects-gone-feature

Levit GS, Hossfeld U (2019) Ernst Haeckel in the history of biology. Curr Biol 29(24):R1276–R1284. https://doi.org/10.1016/j.cub.2019.10.064

Maes J, Teller A, Erhard M, Conde S, Vallecillo S, Barredo JI et al (2020) Mapping and assessment of ecosystems and their services: an EU wide ecosystem assessment in support of the EU biodiversity strategy. EUR

Maes J, Teller A, Erhard M, Conde S, Vallecillo Rodriguez S, Barredo Cano JI, Paracchini M, Malak DA, Trombetti M, Vigiak O, Zulian G, Addamo A, Grizzetti B, Somma F, Hagyo A, Vogt P, Polce C, Jones A, Carré A, Hauser R (2021) EU ecosystem assessment, EUR 30599 EN, Publications Office of the European Union, Luxembourg, 2021, ISBN 978-92-76-30614-6, https://doi.org/10.2760/846428, JRC123783

Meadows DH, Goldsmith EI, Meadow P (1972) The limits to growth, vol 381. Earth Island Limited, London

Moreno CE, Calderón-Patrón JM, Martín-Regalado N, Martínez-Falcón AP, Ortega-Martínez IJ, Rios-Díaz CL, Rosas F (2018) Measuring species diversity in the tropics: a review of methodological approaches and framework for future studies. Biotropica 50(6):929–941. https://doi.org/10.1111/btp.12607

Pyšek P (1998) Alien and native species in central European urban floras: a quantitative comparison. J Biogeogr 25(1):155–163. https://doi.org/10.1046/j.1365-2699.1998.251177.x

Reimchen TE (2000) Some ecological and evolutionary aspects of bear-salmon interactions in coastal British Columbia. Can J Zool 78:448–457. https://doi.org/10.1139/z99-232

Rodhe H, Charlson R, Crawford E (1997) Svante Arrhenius and the greenhouse effect. Ambio 26(1):2–5

Sánchez-Bayo F, Wyckhuys KA (2019) Worldwide decline of the entomofauna: a review of its drivers. Biol Conserv 232:8–27. https://doi.org/10.1016/j.biocon.2019.01.020

Spehn EM, Joshi J, Schmid B, Diemer M, Körner C (2000) Above-ground resource use increases with plant species richness in experimental grassland ecosystems. Funct Ecol 14(3):326–337. https://doi.org/10.1046/j.1365-2435.2000.00437.x

Tilman D, Isbell F, Cowles JM (2014) Biodiversity and ecosystem functioning. Annu Rev Ecol Evol Syst 45:471–493. https://doi.org/10.1146/annurev-ecolsys-120213-091917

Tuomisto H (2010) A diversity of beta diversities: straightening up a concept gone awry. Part 1. Defining beta diversity as a function of alpha and gamma diversity. Ecography 33(1):2–22. https://doi.org/10.1111/j.1600-0587.2009.05880.x

Turner WR, Brandon K, Brooks TM, Costanza R, Da Fonseca GA, Portela R (2007) Global conservation of biodiversity and ecosystem services. Bioscience 57(10):868–873. https://doi.org/10.1641/B571009

Van Klink R, Bowler DE, Gongalsky KB, Swengel AB, Gentile A, Chase JM (2020) Meta-analysis reveals declines in terrestrial but increases in freshwater insect abundances. Science 368(6489): 417–420. https://doi.org/10.1126/science.aax9931

Willson MF, Halupka KC (1995) Anadromous fish as keystone species in vertebrate communities. Conserv Biol 9:489–497. https://doi.org/10.1046/j.1523-1739.1995.09030489.x

Wulf A (2015) The invention of nature: the adventures of Alexander von Humboldt, the lost hero of science. John Murray (Hodder & Stoughton), 2015. xix+472 pp. ISBN 978-1-84854-898-5

Chapter 4
Sustainability: A Behavioral Perspective

Francesco Di Nocera and Giorgia Tempestini

Abstract Human behavior plays a significant role in the deterioration of the environment, contributing to issues such as climate change and biodiversity loss. To address these challenges, it is crucial to comprehend the factors that influence individual choices concerning resource consumption, waste disposal, sustainable transportation, and environmental advocacy. Sustainable behaviors often come with economic, effort, and time costs, so understanding the motivations and incentives that drive individuals' decisions is necessary. The Behavior-Based Design approach emerges as a promising framework for modifying behaviors to benefit individuals, society, and the environment. By targeting behavior change, this approach seeks to promote sustainable practices and mitigate the negative impacts caused by human activities. In line with this, researchers should adopt a structured methodological approach to behavior change within sustainable living. This entails developing interventions to encourage sustainable behaviors and evaluating their long-term effectiveness. By implementing systematic and rigorous evaluation methods, researchers can gain valuable insights into the strategies and techniques that prove most successful in promoting sustainable behaviors over extended periods. Such an approach will provide a deeper understanding of behavior change dynamics and help inform the design of effective interventions to foster sustainable practices.

Keywords Behavior-based design · Sustainable behaviors · Environment · Society

4.1 Introduction

There is a widely accepted scientific consensus that human behavior has significantly contributed to the deterioration of the environment, including climate change and biodiversity loss (Gelino et al. 2021). While it is difficult to pinpoint a single behavior as the sole cause of these complex issues, various human behaviors have

F. Di Nocera (✉) · G. Tempestini
Department of Planning, Design, and Technology of Architecture, Sapienza University of Rome, Rome, Italy
e-mail: francesco.dinocera@uniroma1.it; giorgia.tempestini@uniroma1.it

F. Dal Falco (ed.), *Beyond the Garden*, Designing Environments,
https://doi.org/10.1007/978-3-031-50567-6_4

altered environmental conditions. To address these problems, it is crucial to understand the factors that influence individuals' choices regarding resource consumption, waste disposal, sustainable transportation, and environmental advocacy.

Engaging in sustainable behaviors often incurs costs in economics, effort, and time. For example, biodegradable products may be more expensive, waste sorting requires additional time and effort, and choosing sustainable modes of transportation might take longer. When discussing sustainable behavior, two essential aspects often conflict and influence individual decisions: the environmental benefits or disadvantages and the individual costs or benefits. Indeed, individuals can be differentiated into "pro-socials," prioritizing sustainable actions, and "pro-selves," prioritizing personal benefits over others (Van Horen et al. 2018).

In general, the literature defines pro-environmental behaviors (PEBs) as actions that benefit the environment (e.g., recycling) or avoid behaviors that harm it (e.g., air travel) (Lange and Dewitte 2019). While considering the consequences of our actions on the environment is crucial, in decision-making, short-term results and immediate benefits to our daily lives take precedence over long-term environmental considerations. It is important to note that something providing immediate benefits may not be favorable for the individual in the long run.

In recent years, the Behavioral Design approach has emerged to study and modify behaviors detrimental to individuals, society, and the environment, aiming to promote beneficial behaviors (Khadilkar and Cash 2020). "Beneficial" refers to behaviors that balance individual, collective, and environmental interests. By analyzing current behavioral patterns, interventions can be planned to replace target behaviors with more desirable ones effectively.

Although this approach is typical of behavior analysis (which will be discussed in the next section), the currently most popular orientation is the so-called nudge theory (Thaler and Sunstein 2008). Nudging involves altering the "choice architecture" to predictably influence behavior without eliminating options or significantly changing economic incentives. An illustrative (albeit anecdotal) example is the sticker of a fly placed into the urinals at Amsterdam-Schiphol airport, which significantly reduced unhygienic behavior (spillage) by gently nudging users toward more desirable actions (aiming at the fly).

The concept of nudging has recently attracted academic and nonacademic audiences. The contributions of behavior analysis to the object of study of "nudge theory" have been largely ignored and rarely acknowledged in the literature (Simon and Tagliabue 2018). Yet, the concept of "choice architecture" (aspects of which alter behavior predictably) is entirely superimposable on that of "contingency": the functional relationship between desired behavior and its context. Both nudge theory and behavior analysis operate from the same point: the ubiquitous influence of environmental events on behavior. Even a "random" environmental disposition can influence behavior in one way or another. In any case, the key distinction between the behavioral approach and nudge theory is that the former focuses on consequences, while the latter focuses on antecedents. Nevertheless, the fact that there is a focus on antecedents does not mean that choices do not produce consequences. Accordingly, we strongly advocate for the use of the term "Behavior-

Based Design" to describe our approach, as it emphasizes the alignment with the fundamental principles of functional analysis of behavior. The term "Behavior-Based Design" underscores the importance of understanding the underlying functions and determinants of behavior in the context of designing interventions for sustainable living.

By adopting the Behavior-Based Design framework, we recognize the significance of analyzing the antecedents, consequences, and environmental factors that shape human behavior. This approach acknowledges that behavior is influenced by a complex interplay of variables, including individual motivations, social norms, environmental cues, and reinforcement contingencies.

Through the lens of Behavior-Based Design, interventions can be specifically tailored to address the unique behavioral challenges associated with sustainable living. By thoroughly examining the functions and contexts of behaviors, interventions can be designed to effectively modify behavior patterns, promote sustainable practices, and ultimately achieve positive environmental outcomes.

Furthermore, by aligning with the principles of the functional analysis of behavior, Behavior-Based Design offers a systematic and evidence-based approach to behavior change. It encourages researchers and practitioners to rigorously evaluate the effectiveness of interventions, allowing for a better understanding of which strategies and techniques are most successful in promoting long-term sustainable behaviors.

In summary, the term "Behavior-Based Design" captures our approach to behavior change for sustainable living, highlighting the importance of understanding behavior functions and utilizing evidence-based strategies. By adopting this framework, we can create interventions that are more targeted, effective, and capable of fostering sustainable practices in individuals and communities.

A significant contribution to the considerations discussed in this chapter stems from the meticulous work conducted by Gelino et al. (2021). In their comprehensive review of the literature, the authors examined the application of behavior analysis to sustainability, shedding light on the importance of this approach in designing targeted interventions that have a reliable and measurable impact on individuals' choices toward sustainable practices.

To provide a comprehensive overview of the topic, Gelino et al. categorized the articles based on the type of intervention employed to promote sustainable behaviors or reduce unsustainable ones. This categorization identified nine distinct categories: incentives, feedback, punishments, prompting and education, response effort, stimulus control, self-monitoring, modeling, and commitment and goal. By organizing the literature this way, the authors aimed to capture the breadth of interventions applied in the field and provide a framework for understanding their effectiveness.

Nevertheless, Gelino et al. (2021) acknowledge that other principles may be worth exploring to determine the most effective strategies. This recognition highlights the evolving nature of research in this area and the need for ongoing exploration and innovation. It encourages future studies to delve deeper into behavior analysis principles and examine their applicability to sustainability more systematically and rigorously.

Building upon Gelino et al.'s (2021) work, this chapter aims to clarify concepts and terms of the behavior analysis. By exploring each aspect in detail and examining the empirical evidence supporting their effectiveness, we can better understand how behavior analysis can inform sustainable practices.

By delving into these topics, we can contribute to the growing body of knowledge on behavior analysis and sustainability. Through a more comprehensive understanding of effective intervention strategies, researchers, policymakers, and practitioners can design and implement behavior change initiatives that have a meaningful and lasting impact on promoting sustainable behaviors and mitigating environmental challenges.

4.2 Reinforcement

To fully comprehend sustainability interventions from a behavioral standpoint, delving into the terminology used and clarifying their distinct meanings is essential. Functional analysis of behavior emphasizes the importance of understanding these terms when studying or designing interventions.

One area of confusion arises from the interchangeable use of the terms "incentive," "reward," and "reinforcement." However, these terms carry different implications and should not be used interchangeably (Cameron and Pierce 1994). An incentive is a stimulus presented to an individual before the behavior occurs, intending to encourage the desired behavior. On the other hand, a reward is a stimulus offered to an individual after the behavior has been performed, typically as a positive consequence. In contrast, reinforcement refers to a contingency that increases the frequency of a behavior. It can involve the presentation or removal of a stimulus following the behavior. Positive reinforcement occurs when a stimulus is presented, while negative reinforcement (distinct from punishment) occurs when a stimulus is removed (Skinner 1963). Unlike a reward, reinforcement always increases the frequency of the preceding behavior. Of course, a reward that is provided contingent to the emission of a behavior and that eventually increases the frequency of that behavior is, by definition, a reinforcer.

Reinforcement is the foundation for many energy-efficiency policy interventions, primarily relying on positive reinforcement. Positive reinforcement, such as social approval or praise, is relatively easy to understand in shaping behavior. However, comprehending how negative reinforcement can foster sustainable behavior is more complex.

In the context of sustainable behavior, negative reinforcement involves increasing pro-environmental behavior by avoiding aversive stimuli. It is crucial to differentiate negative reinforcement from punishment, as these terms are often confused. Reinforcement and punishment yield opposite outcomes: increasing and decreasing the likelihood of a response, respectively. In negative reinforcement, a beneficial stimulus is subtracted, resulting in an increased response probability, unlike punishment. To illustrate this in the context of sustainable mobility behaviors, if the targeted behavior is adopting sustainable transportation, it can be reinforced by avoiding punishment or social disapproval.

For instance, a study by Tempestini and Di Nocera (2023) aimed to explore the effectiveness of negative reinforcement in guiding individuals toward sustainable mobility behaviors. The hypothesis posited that introducing an aversive stimulus, such as a delay, associated with unsustainable modes of transportation would act as a negative reinforcer, encouraging the choice of alternative and more sustainable means of transportation. To investigate this, the authors developed a board game where participants aimed to reach 100 destinations in the shortest possible time. The game offered four options: driving, public transportation, cycling, and walking. The player who reached all destinations in the shortest cumulative time determined the winner. After selecting a vehicle, each participant received a "scenario card" indicating the presence or absence of an unexpected event that could increase travel time if driving was chosen. Consequently, sustainable behavior was negatively reinforced by avoiding the less beneficial driving option. Despite driving being the quickest mode of transportation, it was chosen in only 25% of cases overall, while cycling was chosen 54% of the time.

During the initial phase of the game, participants opted for cycling to avoid time increases caused by the unexpected cards. As the game progressed, behavior stabilized, revealing a more realistic pattern of choices. On average, sustainable behavior was negatively reinforced in 80% of cases. This resulted in a significant increase in the choice of cycling over driving due to the presentation of negative reinforcement during the game.

The study by Tempestini and Di Nocera exemplifies how negative reinforcement can effectively influence individuals' decision-making toward sustainable mobility behaviors. By introducing an aversive stimulus associated with unsustainable transportation options, participants were motivated to select alternative and more sustainable modes of transportation. This demonstrates the potential of negative reinforcement for promoting sustainable behavior change.

In conclusion, understanding the nuances of terminology is crucial when studying behavior change interventions in the context of sustainability.

Negative reinforcement encompasses more than just the phenomenon of loss aversion in behavioral economics, which refers to individuals' tendency to avoid unexpected changes rather than pursuing advantages. To delve deeper into this concept, Nicolson et al. (2017) conducted a pioneering study investigating loss aversion among energy bill payers through an online survey. The findings shed light on the participants' preferences, revealing that 93% preferred to avoid switching to a time-based energy tariff, despite the potential financial savings it could offer. The participants' inclination to avoid this change stemmed from their focus on avoiding the unexpected, such as occasional payment increases due to higher consumption, and their desire to minimize financial losses rather than prioritize saving money. Loss aversion, therefore, contributes to greater resistance to change.

It is essential to recognize that loss aversion extends beyond the realm of energy bill payment and can have broader implications. This cognitive bias impacts people's willingness to alter their current situations and makes them less inclined to take risks or embrace new perspectives. Whether it's making career choices, navigating personal relationships, making health-related decisions, or considering environmental matters, individuals with a propensity for loss aversion may be more resistant to change.

However, it is crucial to emphasize that loss aversion does not necessarily result in complete resistance to change. Individuals can overcome this tendency when they perceive that the potential benefits outweigh the possible losses or feel motivated and confident enough to embark on new paths. While loss aversion may make people more hesitant to change, it should not be regarded as an inherent and unchangeable characteristic. With proper awareness of the potential benefits and effective communication, even the most loss-averse individuals can be encouraged to explore new life directions and embrace change.

Understanding the influence of loss aversion on decision-making and behavior can inform the design of interventions and strategies to promote sustainable behaviors. Addressing individuals' concerns about potential losses and emphasizing the benefits and positive outcomes associated with sustainable choices makes it possible to mitigate the impact of loss aversion and motivate individuals to adopt more sustainable lifestyles.

4.3 Punishment

When promoting sustainable behaviors, punishment is often employed in the form of direct fines and taxes or indirect consequences, such as higher consumer prices resulting from industry taxes. While these measures, such as carbon taxes, can effectively reduce consumption, there are several important considerations to keep in mind regarding the effectiveness of punitive interventions.

First and foremost, it is crucial to recognize that punishment does not inherently encourage the development of new desirable behaviors. Punishment involves the presentation of a stimulus (positive punishment) or removing a stimulus (negative punishment), which leads to a decrease in the targeted behavior. However, for punishment to be effective, it must meet specific criteria outlined by Pierce and Cheney (2017): it should be introduced immediately after the undesired response, consistently applied every time the behavior occurs and delivered at a sufficiently high intensity to effectively suppress the behavior. Even when these criteria are met, and the punished response is initially suppressed, it is important to note that punishment only offers temporary suppression. Over time, when the organism encounters the reinforcer that initially increased the frequency of the behavior, there is a potential for the behavior to rebound and return to its pre-punishment level. This illustrates the limited and temporary effectiveness of punishment in behavior change.

A study conducted by Agras et al. (1980) examined the impact of fines on water consumption during the California drought from 1976 to 1978 and assessed their effectiveness in promoting conservation efforts. The results of the study were mixed.

While water use was slightly reduced when fines were implemented, consumption levels returned to near baseline in the subsequent months. From a behavior-analytic perspective, this outcome aligns with the expectations of punishment. As mentioned earlier, punishment must be delivered immediately and consistently after each response that is intended to be punished. In real-world contexts, it becomes challenging to apply punishment functionally since it is not feasible to monitor and control every individual's actions outside of controlled laboratory settings.

Given the limitations and challenges associated with punishment as a sustainable behavior change strategy, alternative approaches should be considered. Instead of solely relying on punitive measures, interventions should focus on positive reinforcement, which involves the presentation of stimuli following desired behaviors acting as reinforces. Positive reinforcement is more likely to facilitate long-lasting behavior change and encourage the development of sustainable habits. By offering incentives, rewards, and benefits contingent to the engagement of sustainable behaviors, individuals are more likely to be motivated to continue and internalize these actions as part of their daily lives.

In summary, while punishment may temporarily impact behavior suppression, there are better approaches for promoting sustainable behaviors due to its limitations and challenges in real-world contexts. To foster lasting change, interventions should prioritize positive reinforcement strategies that provide useful consequences for sustainable actions. By shifting the focus toward positive reinforcement, it becomes possible to create a supportive environment that encourages and sustains pro-environmental behaviors.

4.4 Feedback

Feedback is crucial in promoting behavior change by providing individuals with information about their past performance and suggestions for improving future performance (Brand et al. 2020; Daniels and Bailey 2014). Two critical aspects of the feedback are timing and accuracy. Timing refers to the delay between task completion and feedback delivery, while accuracy pertains to the specificity of the feedback provided to participants (Aljadeff-Abergel et al. 2017). Numerous studies have explored the impact of feedback on electricity consumption, shedding light on its effectiveness.

Hayes and Cone (1981) conducted a study to examine the effects of monthly feedback on reducing electricity usage. Participants received a letter 1 week after the monthly meter reading, separate from their electricity bill, which indicated the percentage change in consumption, the number of kilowatts used, and the actual dollar amount. The results demonstrated that participants who received monthly feedback achieved energy savings, with a 4.7% reduction from the baseline. This study highlights the positive impact of providing individuals periodic feedback on their electricity consumption, enabling them to make informed decisions and adjust their behavior accordingly.

In another study, Kohlenberg and colleagues (1976) investigated how the presentation of feedback could influence energy consumption behaviors in three households. The researchers implemented a system where a light bulb would turn on every time a predefined energy consumption threshold was exceeded, serving as feedback. The results showed that presenting such feedback effectively reduced excessive energy usage in each household. This suggests that immediate and tangible feedback can create awareness and motivate individuals to modify their behavior, leading to more sustainable energy consumption practices.

In a different study, Tiefenbech et al. (2018) employed real-time feedback to guide individuals' water consumption during showering. This feedback directly provided information that made water consumption salient and allowed subjects to immediately modify their behavior. The results revealed a significant 22% reduction in water consumption compared to participants' average consumption per shower. This reduction occurred immediately and remained consistent throughout the experiment, indicating that real-time feedback can have a lasting impact on behavior change.

The growing emphasis on environment-related activities and the increased availability of technologies have contributed to the rising interest in eco-feedback technologies. Often viewed as an extension of persuasive technology research, these technologies aim to address the "environmental literacy gap" by automatically detecting individuals' daily behaviors, such as commuting or showering and providing related information through automated means (Froehlich et al. 2010). By leveraging technologies like computers, tablets, smartphones, and smartwatches, eco-feedback interventions can enhance individuals' awareness and understanding of how their daily actions impact the environment. This increased awareness can motivate individuals to adopt more sustainable behaviors and contribute to positive environmental change.

In summary, feedback, mainly when delivered with appropriate timing and accuracy, has shown to be effective in promoting behavior change in the context of electricity consumption. Studies have demonstrated that providing feedback, whether in monthly reports or real-time notifications, can significantly reduce energy usage. Furthermore, the emergence of eco-feedback technologies offers innovative opportunities to bridge the gap between individuals' behaviors and their environmental impact, fostering greater environmental literacy and encouraging sustainable practices.

4.5 Shaping

The shaping method, as described by Skinner (1975), is a behavioral technique that involves reinforcing increasingly closer approximations to the final desired performance. When implementing a shaping intervention, it is crucial to consider the animal's repertoire, which encompasses the behaviors it is capable of exhibiting based on its species and environmental history (Pierce and Cheney 2017). Shaping becomes particularly useful when you aim to train an animal to perform a response it currently does not exhibit, such as activating a switch. If the baseline observation indicates that the animal does not emit this response (i.e., the operant level is zero), shaping can be employed to establish the desired behavior.

To illustrate the application of the shaping technique in promoting desirable behaviors, let's consider the case of "Motion in Seattle," a program designed to increase sustainable transportation usage on a weekly basis. At the outset, participants were asked a series of questions to assess their starting point, providing insights into their initial repertoire of travel-related behaviors. As an initial incentive, participants were given free public transportation cards to encourage their engagement with sustainable transportation options.

Following the initial assessment, customized programs were created for each participant, taking into account their unique starting point and goals. The shaping process involved progressively reinforcing participants as they achieved specific milestones aligned with sustainable transportation practices. Rewards and congratulations were provided to participants as they made incremental progress toward their goals. This approach allowed individuals to experience a sense of accomplishment and motivated them to continue improving their sustainable transportation behaviors.

Implementing a shaping program with precise customization certainly adds complexity, as it requires careful tailoring to each participant's specific needs and abilities. However, the use of real-time feedback mechanisms can significantly enhance the effectiveness of such programs. By providing immediate feedback that automatically reinforces and encourages improved behaviors, participants can receive timely reinforcement for their efforts and stay motivated throughout their journey of behavior change. This real-time feedback can come in various forms, such as personalized progress updates, reminders, or virtual rewards, further enhancing the shaping process and promoting sustained behavior change.

In summary, the shaping technique is a valuable approach for training animals and promoting behavior change in various contexts. When applied to human behaviors, such as encouraging sustainable transportation choices, shaping programs can be customized to individuals' starting points and gradually reinforce desired behaviors through rewards and feedback. Although implementing such programs with precise customization may be complex, leveraging real-time feedback mechanisms can enhance their effectiveness by providing immediate reinforcement and encouragement to individuals striving for behavior change.

4.6 Gamification

Gamification, a strategy rooted in incorporating gamelike elements into non-game contexts, has proven to be a powerful tool for facilitating learning objectives and motivating behavior change (Deterding et al. 2011). By leveraging gaming dynamics and mechanics, such as earning points, advancing levels, receiving rewards, and earning badges, gamification enhances personal engagement and encourages individuals to achieve goals and overcome challenges. It has found particular application in promoting sustainable behaviors, where it helps bridge information gaps, enhance learning, and motivate individuals to adopt more environmentally friendly practices.

A wide range of products and interventions have been developed based on gamification principles to promote sustainability. These include team competitions, digital games, smartphone applications, data collection tools, and board games (for comprehensive reviews on sustainability games and applications, refer to Douglas and Brauer 2021). By infusing sustainability-related activities with elements of gamification, these interventions create an immersive and interactive experience that captures individuals' attention and drives their engagement.

One example of applying gamification techniques to encourage sustainable behaviors is the work of Merugu et al. (2009). They developed an incentive program called INSTANT (Infosys-Stanford Traffic) with the aim of reducing commute times, enhancing travel comfort, and addressing issues such as congestion, fuel consumption, and pollution. The program leveraged gamification elements to incentivize commuters to travel during less congested periods.

In the INSTANT program, commuters earned credits based on their arrival time at their destination. These credits could be exchanged for monetary rewards, creating a system where the more credits a commuter accumulated, the greater their chance of winning a reward and receiving a higher monetary amount. By providing tangible incentives tied to specific desired behaviors, the researchers sought to motivate commuters to adjust their travel schedules and avoid peak congestion times.

Analysis of the program's data revealed a significant increase in commuters arriving at recommended times to avoid road congestion. The number of commuters arriving during each time slot nearly doubled, indicating the effectiveness of the gamified incentive program in influencing behavior. By incorporating gamification elements and rewarding desired actions, Merugu et al. successfully motivated commuters to alter their travel behavior, resulting in reduced congestion, improved travel comfort, and potential environmental benefits.

This example highlights how gamification can effectively drive behavior change by tapping into individuals' intrinsic motivation and desire for rewards and recognition. By designing interventions that incorporate gamelike elements and provide incentives, gamification strategies can encourage sustainable behaviors and contribute to addressing pressing environmental challenges.

In summary, gamification is a powerful approach for promoting positive behaviors and enhancing engagement in various contexts, including sustainability. By infusing non-game activities with gaming dynamics and mechanics, individuals can be motivated to achieve goals and overcome challenges. The INSTANT program exemplifies how gamification techniques can be applied to influence commuter behavior and bring about meaningful changes in areas such as congestion reduction and environmental impact. Through gamification, individuals are encouraged to actively participate, learn, and adopt more sustainable practices, ultimately contributing to a more environmentally conscious society.

Gamification has gained popularity in the literature, but its underlying dynamics have long been recognized and utilized under the name "token economy." The token economy is characterized by delivering symbolic reinforcements or tokens, such as badges or points, following desirable behaviors. These tokens, which typically have no intrinsic value, serve as conditional reinforcers that can be exchanged for backup reinforcers such as special privileges, food, or activities. Ivy et al. (2017) conducted a literature review and identified six key procedural components of the token economy: (a) target behavior(s), (b) tokens as conditional reinforcers, (c) backup reinforcers, (d) token production schedule, (e) exchange production schedule, and (f) token exchange schedule.

While both gamification and the token economy aim to promote desirable behaviors, they differ in terms of their methodological rigor. The token economy employs a more systematic approach, requiring meticulous and complex planning. It involves the interplay of three interconnected reinforcement schedules: token production (governing when tokens will be delivered), exchange production (governing when tokens will be exchanged), and the cost (in tokens) associated with the available goods or services for exchange. In contrast, gamification does not necessarily require that rewards obtained be compulsorily exchanged for a final reinforcement, as in the token economy.

Despite these differences, both approaches offer valuable tools for behavior change interventions. The token economy provides a more structured and regulated framework for promoting behavior change, emphasizing specific components and reinforcement schedules. On the other hand, gamification offers more flexibility and may focus on immediate rewards without requiring a final exchange. Understanding each methodology's nuances and potential benefits can inform the design and implementation of effective behavior change strategies in various contexts.

4.7 Delay Discounting

The tendency to prioritize immediate benefits over long-term well-being or environmental benefits is a common phenomenon that has been widely studied. This phenomenon, known as "delay discounting," refers to the inclination to choose a smaller but immediate reward over a larger but delayed reward, often resulting in behaviors that are not conducive to long-term goals (Odum 2011). A classic example of delay discounting can be observed when individuals choose between maintaining a healthy body weight and indulging in a piece of cake. Despite recognizing the importance of maintaining a healthy weight, many individuals succumb to the immediate pleasure of consuming the cake. This preference for immediate gratification poses a challenge when the effort and time required to achieve and sustain a healthy weight are compared to the immediate availability of the cake.

The issue of delay discounting is not limited to personal well-being but also extends to sustainability concerns. Recent research has highlighted the devaluation of future air quality outcomes as a potential barrier to engaging in long-term sustainable behaviors, such as reducing private car usage to mitigate emissions (Berry et al. 2017). While reducing greenhouse gas emissions offers immediate benefits to only a few individuals, the long-term consequences of unchecked emissions have far-reaching impacts on entire populations for generations to come. This presents a critical question: How can individuals be motivated to prioritize long-term benefits over immediate rewards?

Addressing this challenge is undoubtedly complex. However, as previously discussed, interventions such as feedback and reinforcement have shown promise in narrowing the gap between individuals' immediate preferences and long-term sustainability goals. By providing individuals with feedback on their energy consumption patterns and reinforcing sustainable behaviors, interventions have successfully encouraged reductions in heating, cooling, and overall energy usage, resulting in tangible energy savings.

Another strategy to tackle the perception of long-term consequences as distant and unattainable is to incorporate initial incentives that motivate individuals to adopt and maintain sustainable behaviors. These incentives can take various forms, such as rewards, recognition, or tangible benefits. By offering these initial incentives, individuals are more likely to initiate their engagement in sustainable practices and develop a sense of momentum and intrinsic motivation over time. With continued reinforcement and support, individuals can gradually internalize the long-term benefits of their sustainable actions, making them more inclined to prioritize these behaviors without immediate rewards.

In summary, addressing the challenge of delay discounting and fostering long-term sustainable behaviors requires a multifaceted approach. Combining interventions that provide feedback, reinforce desired behaviors, and offer initial incentives can help bridge the gap between immediate preferences and long-term goals. By leveraging these strategies, individuals can be encouraged to make choices that prioritize both their long-term well-being and the sustainability of the environment.

4.8 Conclusion

Behavior analysis offers a practical and insightful approach to understanding the mechanisms of behavior change that can foster sustainable living. While the studies mentioned above provide valuable insights, it is essential to note that they represent only a fraction of the research conducted in this area. There is still a need to explore a broader range of ecologically relevant behaviors and systematically identify those with the most significant potential for promoting sustainable change.

To advance the field, researchers should adopt a more structured methodological approach to behavior change within the context of sustainable living. This entails designing interventions and evaluating their effectiveness in facilitating long-term behavior change maintenance. By implementing systematic and rigorous evaluation methods, researchers can better understand which strategies and techniques are most successful in promoting sustainable behaviors over an extended period.

Furthermore, combining the abovementioned interventions can lead to even greater effectiveness in promoting behavior change. An illustrative study by Bekker et al. (2010) focused on electricity consumption behavior among college students living in housing where electricity was included in the rent without individual consumption charges. The researchers established a baseline condition by monitoring daily electric meter readings. Subsequently, they implemented a multifaceted intervention that utilized feedback, incentives, and education to encourage reduced energy consumption. Large posters were prominently displayed throughout the student housing hall, informing residents that they could earn rewards by saving electricity. The posters emphasized the positive impact of their behavior on the environment and specifically within their hall. Additionally, 89 locations within the hall featured electricity-saving tips to further reinforce sustainable behaviors. As a form of feedback, an "electricity-saving thermometer" was employed to provide residents with daily updates on their electricity usage and progress toward their incentive goals. The results of the study demonstrated the effectiveness of the intervention. In the hall where the intervention was implemented, residents achieved substantial energy savings during daytime (16.2%) and nighttime (10.7%) compared to the baseline period. In contrast, the hall without the intervention exhibited lower savings, with residents achieving average reductions of 3.8% during the daytime and 6.53% at night over the same period compared to their baseline situation.

That study exemplifies the potential impact of combining multiple intervention strategies to enhance the effectiveness of behavior change efforts. By utilizing a combination of feedback, incentives, and education, the researchers were able to motivate residents to make significant energy-saving choices. Such comprehensive approaches can be instrumental in promoting sustainable behaviors across various contexts and serve as a model for future research and interventions.

Behavior analysis provides valuable insights into behavior change mechanisms for promoting sustainable living. To advance the field, researchers should systematically identify impactful behaviors for change and adopt a structured methodological approach to evaluating the long-term effectiveness of behavior change interventions. By combining various intervention strategies, such as feedback, incentives, and education, researchers can maximize the effectiveness of their efforts and contribute to promoting sustainable behaviors in diverse settings.

References

Agras WS, Rolf GJ, Lebedeck M (1980) The California drought: a quasi-experimental analysis of social policy. J Appl Behav Anal 13(4):561–570

Aljadeff-Abergel E, Peterson SM, Wiskirchen RR, Hagen KK, Cole ML (2017) Evaluating the temporal location of feedback: providing feedback following performance vs. prior to performance. J Organ Behav Manag 37(2):171–195

Bekker MJ, Cumming TD, Osborne NK, Bruining AM, McClean JI, Leland LS Jr (2010) Encouraging electricity savings in a university residential hall through a combination of feedback, visual prompts, and incentives. J Appl Behav Anal 43(2):327–331

Berry MS, Nickerson NP, Odum AL (2017) Delay discounting as an index of sustainable behavior: devaluation of future air quality and implications for public health. Int J Environ Res Public Health 14(9):997

Brand D, Novak MD, Di Gennaro Reed FD, Tortolero SA (2020) Examining the effects of feedback accuracy and timing on skill acquisition. J Organ Behav Manag 40(1–2):3–18

Cameron J, Pierce WD (1994) Reinforcement, reward, and intrinsic motivation: a meta-analysis. Rev Educ Res 64(3):363–423

Daniels AC, Bailey JS (2014) Performance management, 5th edn. Performance Management

Deterding S, Dixon D, Khaled R, Nacke L (2011, September) From game design elements to gamefulness: defining "gamification". In: Proceedings of the 15th international academic MindTrek conference: envisioning future media environments, pp 9–15

Douglas BD, Brauer M (2021) Gamification to prevent climate change: a review of games and apps for sustainability. Curr Opin Psychol 42:89–94

Froehlich J, Findlater L, Landay J (2010, April) The design of eco-feedback technology. In: Proceedings of the SIGCHI conference on human factors in computing systems, pp 1999–2008

Gelino BW, Erath TG, Reed DD (2021) Going green: a systematic review of pro environmental empirical research in behavior analysis. Behav Soc Issues 30(1):587611

Hayes SC, Cone JD (1981) Reduction of residential consumption of electricity through simple monthly feedback. J Appl Behav Anal 14(1):81–88

Ivy JW, Meindl JN, Overley E, Robson KM (2017) Token economy: a systematic review of procedural descriptions. Behav Modif 41(5):708–737

Khadilkar PR, Cash P (2020) Understanding behavioural design: barriers and enablers. J Eng Des 31(10):508–529

Kohlenberg R, Phillips T, Proctor W (1976) A behavioral analysis of peaking in residential electrical-energy consumers. J Appl Behav Anal 9(1):13–18

Lange F, Dewitte S (2019) Measuring pro-environmental behavior: review and recommendations. J Environ Psychol 63:92–100

Merugu D, Prabhakar BS, Rama N (2009, July) An incentive mechanism for decongesting the roads: a pilot program in Bangalore. In: Proceedings of the of ACM NetEcon workshop

Nicolson M, Huebner G, Shipworth D (2017) Are consumers willing to switch to smart time of use electricity tariffs? The importance of loss-aversion and electric vehicle ownership. Energy Res Soc Sci 23:82–96

Odum AL (2011) Delay discounting: i'm a k, you're a k. J Exp Anal Behav 96(3):427–439

Pierce WD, Cheney CD (2017) Behavior analysis and learning: a biobehavioral approach, 6th edn. Routledge, New York

Simon C, Tagliabue M (2018) Feeding the behavioral revolution: contributions of behavior analysis to nudging and vice versa. J Behav Econ Policy 2(1):91–97

Skinner BF (1963) Operant behavior. Am Psychol 18:503–515

Skinner BF (1975) The shaping of phylogenic behavior. J Exp Anal Behav 24(1):117–120

Tempestini G, Di Nocera F (2023) Promoting sustainable mobility by behaviour-based design: a pilot study. In: Proceedings of the human factors and ergonomics society Europe chapter 2023 annual conference. Retrieved October 2023, from http://hfes-europe.org

Thaler RH, Sunstein CR (2008) Nudge: improving decisions about health, wealth, and happiness. Yale University Press, New Haven

Tiefenbeck V, Goette L, Degen K, Tasic V, Fleisch E, Lalive R, Staake T (2018) Overcoming salience bias: how real-time feedback fosters resource conservation. Manag Sci 64(3):1458–1476

Van Horen F, Van der Wal A, Grinstein A (2018) Green, greener, greenest: can competition increase sustainable behavior? J Environ Psychol 59:16–25

Chapter 5
The Notion of Green Heritage and Integrated Programs of Sustainable Urban Regeneration: The Interplay Between General/Public and Individual Interests

Maria Chiara Romano

Abstract This chapter aims to make a preliminary conceptual cleaning up of the elusive notion of "green heritage" as scattered throughout normative texts (supranational, constitutional, state, regional, and local) that use it for different purposes (urbanistic, aesthetic, cultural, social, historical, of identity). The introduction in Article 9 of the Italian Constitution of the principles of environmental protection, biodiversity, and ecosystems has brought about a paradigm shift according to a twofold order of reasons: the first due to their inclusion among the fundamental principles, which gives them the power to conform the contents of the legislative, regulatory and administrative function, and the second related to their being aimed at the protection of future generations. In this scenario, the issue of the enhancement and redevelopment of urban green spaces is addressed not only from the perspective of the ecological transition initiated by international (UN Agenda 2030) and European (European Green Deal and Next Generation EU) law but also as an effective guarantee of the individual's right to live in healthy and sustainable urban settings. Pandemic and post-pandemic experience has recorded the beneficial effects that the relationship with green areas can produce on the psychophysical well-being of citizens. The current energy and climate crisis highlights the instrumentality of green spaces to decarbonization goals. Within this framework, the new integrated urban and environmental regeneration programs/interventions will be examined. These, overcoming the previous sectorial connotation of land-government policies, are marked by a holistic dimension capable of developing a new administrative function of urban regeneration, aimed at the simultaneous protection of general (ecosystems, biodiversity) and individual (health, psychophysical well-being), as well as material (redevelopment, enhancement, fruition) and intangible (cultural, educational) interests. The essay also aims to analyze a number

M. C. Romano (✉)
Department of Planning, Design, and Technology of Architecture, Sapienza University of Rome, Rome, Italy
e-mail: mariachiara.romano@uniroma1.it

F. Dal Falco (ed.), *Beyond the Garden*, Designing Environments,
https://doi.org/10.1007/978-3-031-50567-6_5

of public and private urban green redevelopment projects that stimulate an effort to adjust the anachronistic category of public property toward a more up-to-date notion of urban green heritage.

Keywords Green heritage · Cultural heritage · Garden · Urban greenery · Urban planning

5.1 Introduction

This chapter aims to reconstruct the notion of green heritage in its plethora of declinations: aesthetic-cultural, landscape, environmental, urban, social, economic. This begs the question of whether it might be possible to identify a holistic notion of green heritage, capable of overcoming the fragmentary and punctiform nature of the numerous different legal frameworks applicable to this area. The mission here is to highlight a trajectory of change in the regulation of green heritage which, starting from a diachronic analysis of the various meanings through which the latter has been taken into consideration by positive law, ultimately bringing about an overarching and all-encompassing function that may be assigned to green heritage, constituting a synthesis of not only individual interests, initially based on a view linked exclusively to private property, but also general/public interests, captured at a transgenerational level that recovers its inevitable collective origin. This synthesizing function could well benefit from the procedural connotation typical of urban planning regulations, which might be capable of harmonizing and facilitating a variety of uses of greenery in Italy. The progressive emergence of functions characterizing green heritage can be outlined in parallel with numerous changes in the role of the State and public powers toward citizens, which, superseding the strongly authoritarian and pan-public law view of the past, now call for innovative forms of shared and cooperative administration, inspired by and reappropriating the solidarity-based principles already inscribed in the provisions of our Constitution.

5.2 Gardens: From Aesthetic-Cultural Value to Landlord/Ownership Privileges to "Building Fever"

The post-unification events relating to the protection of urban green areas, and more precisely of gardens, are particularly telling in that they show how the "inviolable" nature assigned to ownership rights, construed under the individualistic conception crystallized in Art. 29 of the Albertine Statute, contributed toward causing the destruction of gardens and historic Renaissance villas in order to indulge the so-called "Umbertine fever" of the late 1800s and encourage uncontrolled building subdivisions (Severini 2009). The absence of a legislative framework for the protection of gardens and, above all, their reconciliation with a model based on landlord/ownership privileges highlights the initial function of green spaces as the exclusive, and excluding, use of owners of real estate. The social and urban changes of the end of the nineteenth century gave rise to a process of "democratization" experienced in several European countries, albeit not during entirely coinciding timeframes, which led to a change in the concept of gardens, which were no longer considered exclusively places for a few privileged owners, but rather as urban spaces intended for enjoyment by the community as a whole. While originally a dominical or ownership-based approach prevailed, which allowed for their exploitation for commercial purposes, a projection of the absolute primacy of private ownership, the attraction of gardens and urban green areas with a view toward protecting the national heritage began to impose itself, along with the troubled assimilation of green heritage into the category of artistic assets. In this regard, in fact, the subjection of "gardens, forests and landscapes" to the protection granted to works of art had been suggested in the Rosadi bill (Law no. 364 of 20 June 1909) and specifically in an amendment proposed by the Florentine deputy Giovanni Rosadi with regard to certain construction works at Villa Borghese, which had been saved from subdivisions through its acquisition by the State in 1901 (Fig. 5.1). However, this amendment was not accepted since the regulation of "natural landscape assets" was

Fig. 5.1 Temple of Diana, Villa Borghese, Rome. (Source: Author's photo)

delegated to a subsequent law on landscape, which was supposed to also include gardens. The result was a transitory conceptual assimilation of gardens to "natural landscape assets," which gave rise to misunderstandings that continued to persist until the enactment of the code of cultural and landscape heritage in 2004 (Legislative Decree n. 42/2004). The so-called Rosadi-Rava Law (Law no. 364 of 20 June 1909) had not subjected the gardens to the protection envisaged for assets having historical and artistic value on account of the unlimited nature of private ownership, which had justified an aversion to imposing restrictions on real estate, and therefore also on gardens, even those of special historical-artistic value. Ownership and the market, liberal ideals, and the theorization of state intervention for purely conservational purposes and for the protection of the public interest therefore represented the ideological and political coordinates along which the legal framework governing gardens developed between the end of the nineteenth century and the early twentieth century. In fact, the optimization of returns on land, the "building fever," the propensity to consider Rome as fertile ground for real estate speculation converged; these are factors that were all to assign to Rome a specific economic vocation, one which was destined to undermine the industrial potential of the capital city and demean its potential as regards the enjoyment of its green spaces.

5.3 The Gradual Return of the Concept of Gardens Under the Notion of Cultural Heritage

The 1909 law had marked the exclusion of gardens from the binding protection envisaged for works of art and had removed them from public protection for reasons attributable to the divergence between movable property and immovable property, with the consequent perception that gardens were tantamount to "natural landscape assets," despite the emergence of reasons suggesting that the issue not be addressed exclusively from the standpoint of merely protecting nature. After all, when it came to gardens, it was the intervention of the man who had designed their layout, which was subject to protection, and it was to such designs that undoubted historic-artistic value was attributed (Severini 2009).

In a different political and cultural climate, which may also be gleaned from the municipal council of Rome's decision to issue its first urban planning instrument in 1909, it was decided to put in place a form of protection separate from that governing natural landscape assets for gardens, villas, and parks as well, since in the case of the latter, it was not so much the beauty of nature that had to be protected, but rather the transformation of the "natural asset" brought about through artistic design and enriched with cultural depth (Carpentieri 2022). It is particularly interesting to note the reference that was made in the presentation of Law No. 688 of 23 June 1912 (the so-called Ricci-Credaro law) to the works carried out within Villa Umberto (now known as Villa Borghese), with the aim of lamenting the absence of any form of protection for the gardens and pointing out the risk that their artistic value could be destroyed if they were abandoned to mindless construction works. Even before a form of protection for gardens and historic villas had been prepared, Villa Borghese had been at the center of a legal case before the Supreme Court concerning the grant to Roman citizens of the so-called *jus deambulandi* or right of public passage when Prince Borghese had decided to sell the villa, thus preventing Roman citizens from continuing to exercise their right of public passage, which had been granted to them up until then (Gambino et al. 2016). In consideration of Prince Borghese's intention to exercise his exclusive right of ownership, the Municipality of Rome obtained from the Supreme Court (Supreme Court March 9, 1887) the recognition of the right of public use of Villa Borghese by the residents of Rome. This was an early example of the function of collective enjoyment served by urban green areas—in this case the right to stroll in a pleasant place—which would only be theorized later, transcending any merely ornamental or decorative meaning.

Through Law No. 688 of 23 June 1912, it was decided to extend the protection that had previously been strictly afforded only to works of art, that is, movable assets, also to immovable property (including "villas, parks, and gardens"), inaugurating the return of gardens under the multifaceted notion of cultural heritage and their subjection to the relevant legal framework, which would continue under subsequent legal frameworks as well (Giannini 1976) (Fig. 5.2).

Fig. 5.2 Venaria Reale, Turin. (Source: https://www.flickr.com/photos/25896590@N04/9342955433)

5.4 The So-Called Bottai Law (Law No. 1089/1939) and the Cultural and Landscape Heritage Code (Legislative Decree N. 42/2004): From Continuity to Prospects for Change

The qualification of gardens as cultural heritage is confirmed in Art. 1 of the so-called Bottai Law No. 1089/1939, which introduces a further element into the legal framework governing public gardens and parks, that is, the automatic qualification of the latter as cultural heritage, analogous to provisions governing works of art and monuments belonging to the public. The equation between public ownership and submission to the specific protection regime envisaged for cultural heritage remained substantially unchanged in the 2004 cultural and landscape heritage code (Legislative Decree no. 42/2004). Under these provisions of law, it is the public ownership of the assets that determines the application of the specific protection regime, in a context which, from a diachronic standpoint, has become useful for the protection of the "public interest," initially construed as an interest coinciding with the interest of the State (according to the standard model of the liberal State of the early 1900s) and later expanded to also include the subjective interests in the territorial bodies comprising the Italian Republic, with the aim of preserving the cultural heritage, which met a predominantly aesthetic criterion. The qualification of gardens as cultural heritage also fits into this same approach, since gardens are perceived as assets capable of expressing that concept of "artistic beauty," still far from including any appreciation of their historic, identity-based, and spiritual potential, as well as the fact that they are meant to be enjoyed by the general public. The prominence given to the ownership regime gave rise to two further implications: on the one hand, the equation of private nonprofit entities and ecclesiastical entities with public bodies, and on the other, the need for an administrative procedure to identify the specific cultural importance of privately owned assets (Sandulli 1954). This contrast between public ownership and private ownership is also maintained in the current cultural and landscape heritage code. For movable and immovable assets owned by the State, which are of artistic, historical, archaeological, or ethno-anthropological interest, a sort of presumption of "cultural value" is assumed, with the consequent precautionary submission to the specific protection regime provided under the cultural heritage code, which can be overcome only following a negative outcome of the administrative verification procedure (Art. 12, legislative decree n. 42/2004). For assets owned by private individuals, on the other hand, the procedure for declaring specific cultural interest is always necessary (Article 13, Legislative Decree No. 42/2004). Such procedure varies, depending upon the specific categories of assets listed in Art. 10, paragraph 3, Legislative Decree no. 42/2004, which determines their subjection to the specific rules of protection and optimization envisaged under the code. Under the cultural heritage and landscape code, gardens, along with villas and parks, are explicitly mentioned in Art. 10, paragraph 4, lett. (f), among immovable assets, which, if publicly owned, are presumed to be of historical or artistic interest, unless the aforementioned verification procedure results in a negative outcome. If instead they are owned by private persons or entities, they must be subjected to a different declaration procedure, which in the case of gardens must verify the existence of a "particularly important" historical and artistic interest, and therefore of a high degree of "cultural value," which would give rise to the imposition of the restriction. Furthermore, gardens can, on a subsidiary basis, that is, if they are not public property or are not of particularly important historical and artistic interest, also be classified as "landscape assets" pursuant to Art. 136, paragraph 1, lett. (b), if they are distinguished "for their uncommon beauty." Gardens are therefore taken into consideration by the Code on the basis of different heterogeneous criteria, which confirm the plethora of functions that they are capable of serving. However, the regulatory qualification that examines them exclusively from the standpoint of their protection as cultural or landscape assets (albeit currently under the broad interpretation formulated since the works by the Franceschini Commission established in 1964, which qualified them as "testimonies having value for civilization") is still narrow. In fact, the provisions of the Code do not sufficiently focus on the element of public interest, meaning not exclusively with regard to the ownership regime and the desire to preserve the asset from a static perspective, but rather delineated in a dynamic and reactive manner, that is, capable of supporting the pursuit of the historically ever-changing interests of the community. Territory, and in this case green territory, can and must be taken into consideration with a view to safeguarding all of the plethora of public/general interests that it is capable of fulfilling, meaning that the legal framework set forth in the cultural heritage and landscape code should not exhaust and/or exclude other legal frameworks. The so-called Charter of Historic Gardens of 1981 (so-called Charter of Florence, signed by the International Committee of Historic Gardens ICOMOS-IFLA), for example, defines "historic parks and gardens" as "living monuments," meriting specific rules, the distinctive feature of which lies in the interplay between artistic creation and natural components (Brocca 2022). It is therefore necessary to systematically coordinate with other similar legal frameworks (urban planning, ecological, botanical, medical, social), which focus on heterogeneous elements and express the multipurpose vocation of green heritage: aesthetic-cultural, identity-based, nature conservation, the fight against climate change, containment of soil consumption, psychophysical well-being, and the enhancement of biodiversity.

5.5 Art. 9 of the Constitution and Its Reform of 2022

In continuity with the two laws of 1939 (no. 1089 and no. 1497), respectively, dedicated to the protection of assets of artistic and historical interest and to natural landscape assets, Article 9 of the Constitution is inspired by the same conceptual distinction between artistic and landscape protection. This dividing line is accompanied, after the constitutional amendment of 2022, by a different approach, which, in adding environmental protection to its fundamental principles, explicitly cites the protection of biodiversity and ecosystems with specific reference to future generations. This innovation, which codified previous guidelines of the Constitutional Court, offers an additional interpretational perspective, which opens the door to considering nature as a "value in and of itself," the inadequate conservation of which could negatively impact future generations. Recovering the very close nexus between the protection of natural capital, green heritage, and human health, the latter is to be seen as a fundamental right of the individual and the interest of society as a whole (Article 32 of the Constitution). The anthropocentric vision initially inherent in environmental law is accompanied by "a unitary conception of the environmental asset including all natural and cultural resources" (Const. Court, judgment no. 210/1987). The reference to biodiversity and natural capital is linked to a new holistic approach developed by the international community following the pandemic crisis and other frequent natural disasters of recent years. This approach summarized in the *One Health* strategy (Fig. 5.3), which had already been formulated in 2004 within the *Wild Conservation Society* through the issuance of the *Manhattan Principles* and was initially used only with respect to animal health, zoonotic epidemics, and antibiotic resistance, has contributed toward a classification of human, animal, and environmental health in unitary and interrelated terms (Scotti 2022). At the center of this new strategy is the scientific assumption that the integrity of the planet affects the health of humans and animals. The weakening of the anthropocentric perspective of exaggerated exploitation of natural resources is accompanied by a

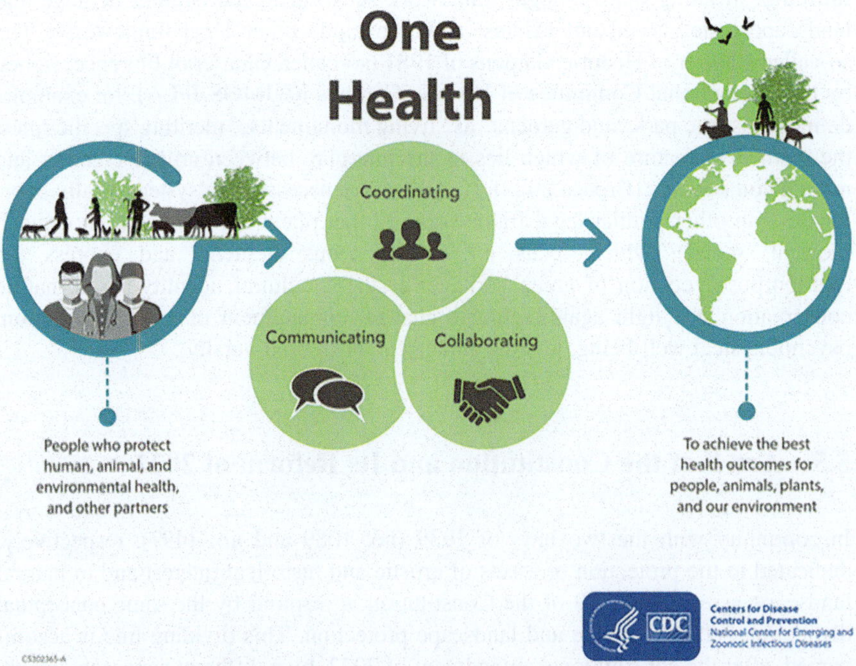

Fig. 5.3 The One Health strategy. (Source: Centers for Disease Control and Prevention, National Center for Emerging and Zoonotic Infectious Diseases (NCEZID) website)

more stringent regulation of economic activity, which goes beyond the concept of sustainable development introduced by the principles of international environmental law and makes economic initiative functional to the pursuit of social and ecological and not merely market-based objectives, focusing on the so-called *Nature Based Solutions (NBS)*, as may be gleaned from the minimum environmental criteria used in the award of public contracts or from the establishment of sustainability reporting for companies (Directive (EU) 2022/2464 of the European Parliament and the Council of December 14, 2022).

The *One Health* strategy also inspired the United Nations 2030 Agenda for Sustainable Development, which, while not citing it explicitly, mentions in its preamble the transversal dimension of human health. The same perspective is also found in the PNRR, in particular with regard to the specific attention dedicated to cities as institutional subjects involved in the fight against climate change and in ecological transition, in which a convergence is seen between the goals of the UN 2030 Agenda and the renewed consideration that the PNRR assigns to urban green areas in pursuit of the well-being of citizens. In fact, urban green areas are expressly mentioned in target 11.7 of Goal 11 of the UN Agenda 2030 relating to "Sustainable cities," with the aim of increasing access to them by local communities.

The international and European regulatory context (*UN Agenda* 2030, *European Green Deal, European Strategy for Biodiversity 2030*) and the projects financed with funds from the *Next Generation EU* also allow us to interpret the PNRR from a *One Health* perspective and, from this viewpoint, have favored a paradigm shift in the function assigned to urban green areas, exceeding the historical-artistic dimension that emerged under the legal framework applicable to gardens (public and private), which still predominantly aestheticizes landscape, in view of the beneficial effects that urban green areas can produce on the reduction of pollution and on the psycho-physical well-being of citizens (Ivancich 2023). In this context, the PNRR attributes a crucial role to the nature protection system, made up of park authorities and marine protected areas (ln 394/1991), which according to the European strategy for biodiversity should cover at least 30% of the earth's surface and land and aqueous surfaces by 2030. Biodiversity monitoring should be facilitated by the use of new satellite technologies capable of preventing hydrogeological risks, desertification phenomena, and alterations of ecosystems and natural habitats. It is no coincidence that ecological transition should benefit from the expansion of digitalization.

The *One Health* strategy has therefore extended to other European and national public policies and forges a new direction in the man-nature relationship expressed in the interpretation of environmental law from a perspective that is no longer anthropocentric but rather eco-centric. Following this approach, without supporting dangerous eco-centric drifts aimed at prophesying the advent of an era defined as "post-humanism" (Peratoner 2009), strategies that are integrated, participatory and inclusive should be put in place, aimed at recovering a man-nature relationship within which the regulation of green heritage not only performs a function serving human health and our psychophysical well-being, at an individualistic level, but also contributes toward establishing the ecological, urban, and social conditions that ensure the preservation of ecosystems, within a broader and more general all-encompassing dimension.

5.6 Urban Greenery and Urban Planning Standards: An Insufficient Perspective

The statistical forecasts that, from a global perspective, measure the unstoppable phenomenon of urbanization estimate that, by 2050, 70% of the world's population will live in cities, with the risk of irreversibly altering the balance between built-up areas and green spaces. In this context, it has become a matter of urgency to identify a legal notion of urban green areas capable of guiding the exercise of planning powers of local administrations, as well as identifying more efficient forms of consultation between the territorial (and non-territorial) entities with planning authorities. The aforementioned codification in Art. 9 of the protection of the environment, biodiversity, and ecosystems, while not entirely innovative, since the principle of environmental protection had already been theorized in the constitutional case law (Carpentieri 2021), is nonetheless certainly appreciable for its positioning among constitutional provisions concerning fundamental principles and therefore for the ability to internally influence the exercise of regulatory and administrative functions and in particular those of territorial planning. Despite the growing focus that urban planning law places on urban green areas, in consideration of the aforementioned interrelationships between green areas and the protection of human health, reduction of pollution, protection of the soil as a nonrenewable ecosystem resource, and the psychophysical well-being of citizens, to date this has not led toa unitary legal definition of urban green areas, with the exception of a reference in ministerial decree no. 1444/1968 on urban planning standards, which resorts to a purely dimensional conception, devoid of positive indications useful for guiding the regulation of urban green areas (Mari 2018). The ratio between public green areas and the extension of built-up areas is one of the indicators proposed for monitoring the achievement of the Sustainable Development Goals (SDGs) within the previously mentioned Goal 11 of UN Agenda 2030 concerning sustainable cities.

The formula used by the aforementioned ministerial decree is partly anachronistic and no longer adequate for the ever-changing social, urban, and economic context, as it refers to a historical moment in which urban greenery was mainly understood as a means to slow down the building expansion of the late sixties. In municipal urban planning regulations, the percentage of urban green areas that meet the parameters of the decree ends up being satisfied by including the collective spaces intended for play and sport, effectively leading to the improper use of the "equipped green areas," which are not always capable of serving the plurality of functions and needs that urban green areas should provide to the community. Administrative case law has taken care of this by correcting the aforementioned interpretation of equipped public green areas, reiterating that "in the areas falling within zone F and intended for equipped public green areas, the green area ... itself constitutes 'public or private equipment for public use' to which such areas are reserved, and the community's enjoyment of green areas is their standard function within the general organization of the municipal territory. [. . .]. Therefore, in the meantime, such equipment is permitted since, with regard to building or architectural characteristics, due to their size or the manner in which they fit into the context, they are compatible with the designated use as public green areas" (Council of State, Section IV, 08.06.2013, No. 4148).

The inputs originating from international and European law have led lawmakers to issue law No.10/2013, the only law in our legal system inspired by a systemic approach aimed at involving not only local authorities but also private individuals in the protection and optimization of urban green areas.

Law No.10/2013 pursues a plethora of objectives focused on promoting a different notion of urban greenery no longer confined to a purely urban planning perspective but capable of developing the potentialities intrinsic in natural heritage, with the aim of interrupting the curve of biodiversity loss, mitigating the macroscopic effects of climate change and satisfying the social demand for green areas in urban centers. Noteworthy aspects of the law include the establishment at the Ministry of the Environment and Energy Security of a committee for urban green areas tasked with monitoring the application of current regulations with the specific "purpose of increasing public and private green areas." The committee's responsibilities include the proposition of a national plan, which, in agreement with the Unified Conference, "sets criteria and guidelines for the creation of permanent green areas around the major conurbations and rows of trees along roadways, to allow for an adaptation of construction and public and educational infrastructures which ensures the requalification of buildings [...] including by enhancing walls and roofs with greenery, creating gardens and vegetable gardens and improving spaces" (Article 3, paragraph 2, letter c., Law no. 20/2013). The establishment of a national plan on urban green areas is certainly an ambitious choice by our lawmakers, which, however, raises some important issues, above all, the issue of the relationship of this new planning instrument with the traditional territorial planning instruments. From the contents of the aforementioned Art. 3, paragraph 2, lett. (c), it is clear that this is a plan essentially falling under the category of urban planning instruments. Indeed, it is not capable of having a conforming effect on property and or on territory but can only guide the substance of territorial planning and in particular that of municipalities and metropolitan cities. The committee, however, has chosen to issue not a plan but a strategy with the aim of specifying criteria and guidelines for the creation of permanent green areas (around the major conurbations) and rows of trees. Uniform criteria and guidelines are thus established at the national level, which must be put into action through the issuance of plans at the most appropriate territorial level.

Art. 6 of Law No. 10/2013 promotes a series of worthy local initiatives all united by the aim of expanding urban green areas (e.g., the creation of green belts around conurbations, the reduction of the impact of construction and making areas subject to new construction greener, plus the creation of large public green areas as part of urban planning). However, being purely optional, no intermediate measures have so far been issued capable of making the initiatives envisaged thereunder mandatory, and this coupled with the absence of control mechanisms or substitutes constitutes one of the main causes of the failure to fully implement the provisions, despite the fact that they are decidedly virtuous. The systematic interpretation of the legal framework leads to the representation of urban green areas as a "permanent planning strategy" (Mari 2018) capable of predetermining and guiding the discretion of planners and of reaffirming the principle of the prohibition of land consumption, meaning a "non-renewable ecosystemic natural resource" (Constitutional Court, no.179/2019).

The partial implementation of the law was followed by municipalities issuing regulations and plans on urban green areas, which present unchanging characteristics. Indeed, while Art. 4, paragraph 2, of law No. 10/2013 had provided for mechanisms to guarantee the effectiveness of ministerial decree No. 1444/1968, providing for the approval of urban planning changes for the municipalities that had not satisfied the minimum quantities of public spaces reserved for public green areas, the poor implementation of this provision, due to the failure to prepare the annual monitoring reports, has not prevented the municipalities from emphasizing in the regulations on urban green areas their identity-based nature for the municipal area as well as their importance for the quality of life of residents (see Regulation of the municipality of Venice for the protection and promotion of green areas in the city; Regulation of public and private green areas in the city of Turin). At the international level (United Nations Convention on Biodiversity) and the European level (EU Biodiversity Strategy 2030, European Commission Communications on Forests, Natural Capital, Green Infrastructure and Land Take, COM (2013) 249), local governments are encouraged to include elements of biodiversity and *Nature Based Solutions (NBS)* in spatial planning. These most certainly include green infrastructure defined as "a network of natural and semi-natural areas capable of providing a broad range of ecosystemic services" (2017 Annual Report of the Committee for the Development of Public Green Spaces). This network of green spaces is capable of producing social and ecological benefits (Fig. 5.4). Europe has promoted the development of a roadmap to promote the greening of EU cities and to maintain green spaces by 2030, inspired by the principle of ecological urbanism and, most importantly, trusting in the quality of urban planning (European Parliament resolution of 17 September 2020 on the European year of green cities 2022, 2021/C 385/20).

From the *National Strategy on Urban Green Areas*, the innovative experiences of municipalities (green regulations and plans), as well as the presentation of projects originating from private entities, it is possible to discern a tendency to abandon a notion of urban green areas construed simplistically from an urban planning standpoint—that is, based on a purely quantitative criterion—in favor of the enhancement of various different functions that green areas serve with regard to specific districts of the city (residential, manufacturing, services, historical interest), thus highlighting the typological richness of urban green areas and their importance for the livability, sustainability, and resilience of cities. Thus, it would seem that the "right to green spaces" can be included within the broader "right to the city" as a fundamental guarantee of urban welfare (Lefebvre 1968; Auby 2013).

Fig. 5.4 Vertical Garden, Milan, Stefano Boeri Architects. (Source: https://www.flickr.com/photos/bulgerhoog/48546770237/in/photostream/)

5.7 Urban Regeneration and Public Contracts Code: Protection of Green Areas, the Interplay Between Public and Private Forces

The growing international, European, national and local importance that, in the post-pandemic era and the climate and energy crisis, urban green spaces hold for the well-being of local communities and for the protection of biodiversity may also be gleaned from a number of regional legal frameworks on urban regeneration. These, in pursuing the objective of lower soil consumption, climate mitigation, reduction of hydrogeological risks, and improvement of the functional, environmental, and landscape quality of territories set for themselves the objective of creating an integrated and multi-sectoral approach, based on programs and intervention projects of high ecological quality. In this context, the legal framework governing green areas, both public and private, can no longer be covered by uncoordinated sector-specific legal frameworks, but rather demands inclusion within a broader and more systemic vision, capable of optimizing and guiding the possible integrations and synergies among various administrative areas of competence, as well as between public and private interests (Di Lascio 2018).

From this perspective, Lombardy regional law No. 18 of 26 November 2019, which under Art. 2 defines urban regeneration as "the coordinated set of urban planning-building interventions and social initiatives which may include the replacement, reuse, redevelopment of the developed environment and the reorganization of the urban layout, not only in relation to the recovery of degraded and abandoned areas, but also through the construction and management of equipment, infrastructures, green spaces and services and the recovery or enhancement of existing ones, with a view to ensuring environmental and social resilience, technological innovation and increasing biodiversity in the urban environment." The tools designed to achieve the aforementioned objectives of urban regeneration referred to under the Lombardy regional legislation explicitly include collaboration between local authorities and park authorities aimed at creating a model of sustainable territorial development, as well as their implementation through the instruments of public-private partnerships and negotiated planning (Di Lascio 2017).

At the national level, the legal framework governing this integrated approach, inspired by cooperation between public administrations and private individuals, is found in the new code of public contracts (Legislative Decree No. 36/2023), which expressly introduces the mechanism of "social partnership," already coined in these terms in the opinion of the Council of State on the outline of the previous code of public contracts. Art. 201, paragraph 1, lett. (a), of the new code, reformulating and merging the provisions of articles 189 and 190 of the 2016 code ("*horizontal subsidiarity interventions*" and "*administrative bartering*") provides that the contracting authorities can establish, through a general document, criteria and conditions for the conclusion of social partnership contracts concerning the management and maintenance of areas reserved for urban public green areas as well as the completion of works of local importance. The provision introduces a negotiating figure, which, through a mutual *datio in solutum* (mutual in-kind exchange), aims to achieve goals of general interest for citizens and for the local community. Citizens, whether individuals or in associations, and small- and medium-sized enterprises, instead of paying a tax debt in cash, fulfill their obligation through the performance of work or service, while the territorial entity, on the other hand, does not pay cash for work or service, but offsets its own tax credit.

This provision is extremely interesting as it constitutes the application of the principles of solidarity and horizontal subsidiarity referred to in Art. 6 of the new public contracts code and already provided under the Constitution, respectively, in Art. 2 and Art. 118, paragraph 4 (Palazzo 2022). Art. 201 involves private persons and entities of the service sector in urban green development strategies by resorting to a unitary notion of "social partnership," meaning a stable form of collaboration between private individuals and the administration for purposes of serving a general interest. The provision, however, still leaves unresolved the coordination with Art. 55 of the services sector code (Legislative Decree no. 117/2017), which regulates the institutes of co-programming and co-planning with exclusive regard to services sector entities, allowing only the latter to participate in the planning phases for planning the needs to be satisfied and planning the interventions to be implemented. Indeed, if Art. 201 aims to foster forms of shared administration between administrations and private individuals and entities (including services sector entities), it remains unclear why the legal framework on social partnership does not also include the guarantee of co-programming and co-planning (Constitutional Court no. 131/2020).

In fact, what forms the basis of social partnership is the link between private subjects/citizens, service sector entities (including micro-, small-, and medium-sized enterprises) and the local territory, which highlights their ability to filter and express the needs of the community of residents and therefore also the social demand for urban green areas. Regardless of the interpretational issues yet to be resolved, it is clear that the exercise of the administrative function of protection and enhancement of urban green areas should not be reserved exclusively to public authorities and to municipal administrations in particular, but should be performed more appropriately by making use of social partnership, which fosters cooperation between public administrations and private individuals, as an expression of the so-called "solidarity society" (Constitutional Court, no.131/2020), in the pursuit of environmental, social, recovery of unused real estate purposes, with the aim of overcoming the financial difficulties faced by public administrations.

5.8 Conclusions

The analysis of the laws and regulations, which in various sectors (the legal framework on cultural and landscape heritage; constitutional provisions that protect the environment, biodiversity, and ecosystems; laws on the development of urban green spaces; ministerial decrees on urban planning standards; regional laws on urban regeneration; the new code of public contracts) take into consideration green heritage, has brought out the plethora of heterogeneous functions that they are capable of serving and, most importantly, has rendered their qualification centered on the ownership regime anachronistic and inadequate: green areas, both public and private, are essential for the well-being of citizens, the conservation of biodiversity, enjoyment as a cultural or landscape asset, the fight against climate change, the containment of soil consumption, and the redevelopment of built-up areas. This multifunctional nature of green heritage is consistent with the social function that private property must express (Art. 42, paragraph 2, Constitution), according to a formula that recognizes the flexibility necessary to adapt to constantly evolving social needs. The task of rendering various different uses of green heritage compatible and coexistent must be entrusted to territorial and urban planning tools (Cerulli Irelli 1985).

However, the Italian legal framework is not yet in line with the indications originating from the international and European context, which require integrated and multi-sectoral approaches aimed at expanding green areas in their many forms, also by fostering urban regeneration interventions, which tend to use, from a unified perspective, both natural and cultural components to achieve the abovementioned positive impacts in terms of the use of cultural heritage, the conservation of biodiversity and the psychophysical well-being of local communities. The multifaceted social and ecosystemic functions that green heritage is capable of offering to the community show that the merely aesthetic or historicist vision has now been superseded, in favor of a different solidarity-based perspective, which should further enhance the initiatives of civil society in carrying out activities of general public interest.

The cause of the malfunctions, inefficiencies, and delays in our legal system with respect to the indications originating from the supranational context can, it seems be more generally ascribed to the complexity and duplications of administrative, territorial and urban planning responsibilities and control measures, which continue to give rise to bureaucratic complications and decision-making deadlocks, attributable to the difficulty of systematically interpreting the various different rules governing green heritage.

Regulatory and procedural simplification could represent an effective antidote leading to the correct enjoyment/optimization of green heritage under the Italian legal system and for the correct balance between general and individual interests (emblematically: on the one hand, protection of the environment, biodiversity, and health and, on the other hand, protection of individual health and private economic initiative), which however, at present—in light of our segmented regulatory framework and fragmented structure of administrative powers—appear destined to lead to ongoing insurmountable conflict.

References

Auby JB (2013) Droit de la ville. Lexisnexis, Paris

Brocca M (2022) PNRR e "verde storico". Aedon 3:141–146. https://doi.org/10.7390/106606

Carpentieri P (2021) Paesaggio, ambiente e transizione ecologica. Giustiziainsiemeit:1–10

Carpentieri P (2022) Paesaggio, urbanistica e ambiente. Alcune riflessioni in occasione del centenario della legge Croce n. 778 del 1922. Giustizia-amministrativa.it, 1–24

Cerulli Irelli V (1985) Pianificazione urbanistica e interessi differenziati. Rivista trimestrale di diritto pubblico 2:386–441

Di Lascio F (2017) Spazi urbani e processi di rigenerazione condivisa. In: Di Lascio F, Giglioni F (eds) La rigenerazione di beni e spazi urbani, Contributo al diritto delle città. il Mulino, Bologna, pp 65–88

Di Lascio F (2018) Quali tendenze in corso nella rigenerazione delle città? Rivista Giuridica dell'Edilizia 2:135–151

Gambino AM, Genovese A, Sirgiovanni B (2016) Beni comuni. Nuovo diritto civile 1(3):301

Giannini MS (1976) I beni culturali. Riv trim dir pubbl 1:3–38

Ivancich V (2023) Noi e l'albero, Perché il contatto con la natura fa bene alla nostra salute. Corbaccio, Milan

Lefebvre H (1968) Le droit à la ville. Anthropos, Paris

Mari G (2018) Il verde pubblico e la pubblicizzazione del verde privato. Rivista giuridica dell'edilizia 1:39–62

Palazzo D (2022) Pubblico e privato nelle attività di interesse generale, Terzo settore e amministrazione condivisa. Giappichelli, Turin

Peratoner A (2009) Quale antropocentrismo? Ripensare la persona umana in relazione all'ambiente. In: Riduzionismo e antiumanesimo. anthropologica.eu, pp 39–53. Retrieved January 1, 2009, from http://www.anthropologica.eu

Sandulli AM (1954) Natura e funzione della notifica e della pubblicità delle cose private di interesse artistico e storico qualificato. Riv trim, dir proc civ:1023

Scotti E (2022) One Health: per un'integrazione tra salute umana e ambientale. In: Bella FA (ed) One Health: la tutela della salute oltre i confini nazionali e disciplinari, Per un approccio olistico alla salute umana, animale e ambientale. Editoriale Scientifica, Naples, pp 45–64

Severini G (2009) I giardini come beni del patrimonio culturale: storia di una legge e questioni interpretative. Aedon 1:1–6

Sitography

Annual report 2017 of the Committee for the Development of Public Green Areas, from https://www.mase.gov.it

Judgments issued by the Constitutional Court, from https://www.cortecostituzionale.it

Judgments issued by the Council of State, from https://www.giustizia-amministrativa.it

Chapter 6
The Anti-garden

Emanuele Quinz

Abstract Crystallising in countless myths the principles of origin and universal mutation, the garden is the place where the rituals of metamorphosis are played out, where vital impulses emerge, where the distinction between the artificial and the natural is blurred, as are the boundaries between the plant, the mineral and the animal. A place of strangeness, licence and whimsy, the garden is defined as a place of 'indistinction' where the kingdoms merge. Starting from the intersection of literature and art history, this contrasting image of the garden is particularly prevalent in movements that turn reason on its head – such as Mannerism and Romanticism, or more recently with the experimental aesthetics of the 1960s or with postmodernism. In its hybrid nature, where physical and sensory immersion is intensified by symbolic stratifications, the garden is experienced as a place open to the disparate, the irregular, the chaotic and the unpredictable – which, beyond representations, allows us to experience an encounter with otherness – and to found an 'art of experience', which never ceases to resonate in contemporary art.

Keywords Garden · Experience · Mannerism · Metamorphosis · Scripted spaces

6.1 The Anti-garden

Outside, all around, there is the forest, black, impenetrable, frightening. The forest is inhabited by wild beasts and haunted by spirits and divinities.

The garden has always been linked to the enclosure: a closed, protected space. Its origins have been identified in the clearing, like an eye opening in the darkness, a crack in the blackness of the woods, pierced by the light. A circumscribed space, where man carves out his territory and establishes a principle of order and reason, in the face of chaos, in the face of the mystery of the non-human. It is also to the clearing that Giambattista Vico returns 'the original deed of appropriation that first

E. Quinz (✉)
Department of Visual Art, Université Paris 8 Vincennes-Saint Denis, Saint-Denis, France
e-mail: equinz@univ-paris8.fr

F. Dal Falco (ed.), *Beyond the Garden*, Designing Environments,
https://doi.org/10.1007/978-3-031-50567-6_6

opened the space of civil society' (Harrison 1992, p. 6), which constitutes the founding of cities, nations and civilisations.

From this starting point, the history of gardens can be read in terms of this polarity, between darkness and light, between enigma and clarity, between the infinite and the finite and between the space of nature and that of man – whether it be the space of civilisation surrounded by nature, or of nature surrounded by the city.

In literature from the Middle Ages to the Baroque and from the lyric poetry of the Stilnovo to the epic poems of chivalry, this polarity is a recurring motif. On the one hand, there is the obscure forest, the reserve of otherness, the place of the fall, of regression into savagery, of the mysteries of sexuality and madness, the place of wandering and unreason, and of the anarchy of matter and the senses, where outcasts, hermits, brigands, fugitives and lovers meet, and, on the other, the garden, the luminous bulwark of a domesticated and ordered nature, the emanation of the principle of identity and non-contradiction. But very quickly, in contrast to this polarity, which may seem too linear, the garden reveals all its complexity, its play of shadows and glints, contrasts and ambiguities.

If, in Dante, the two extremes appear in a progression that leads from the *selva oscura* in which the poet gets lost, through Hell, where other even darker forests extend, like that of the suicides, to the garden of Eden at the point of Purgatory, to finally reach ultimate salvation, in Tasso's *Giardino di Armida*, or in the series of sumptuous, tempting parks through which the protagonist of Giambattista Marino's Baroque poem *L'Adone* (1623) passes, the boundaries become blurred, more and more lures are hiding in the gardens, and enchantment becomes synonymous with illusion.

In all these stories, walking through the forest or the garden is an intense, shattering experience that brings about a change, a transformation in the subject. The opposition between forest and garden embodies the opposition between loss and self-conquest, between *errare* (wandering, physically, but also raving, straying from the right path, making errors) and knowing (oneself).

Still heavily steeped in mythology, all these stories paved the way for the Bildungsroman that was to become the preferred narrative mode of Romanticism, in which a series of twists and turns punctuate the existential journey through which the individual forges himself, first in relation to nature and the non-human and then to society and his peers.

From Goethe's botanical gardens, where he searches for the *Urpflanze* principle and plays out his *Elective Affinities*, we move on to Novalis's search for the 'blue flower', to Tieck's fairy tales, where, always beyond the confines of the city, the characters keep getting lost, in forests that resemble gardens and in gardens that resemble forests and in a nature shrouded in mist and mystery. Increasingly, natural places appear as the antithesis, at once intoxicating and painful, of prosaic everyday urban reality, as places of enigmatic visions and apparitions and of transfiguration and transformation, where the emotion of nature is defined as a form of ecstatic nostalgia (*Sehnsucht*) and where getting lost means rediscovering memories of a deep-rooted elsewhere gone forever.

All these stories reveal that the garden is always doubled by an anti-garden, defined by an opposition, the phantasmatic evocation of what it is not. This double persists like a haunting: on the one hand, the garden encourages us to escape into its luminous alleys, its shady corners, to let ourselves be surrounded and reassured by its silences charged with life, and in its vibrations of colours and smells; on the other, it confronts us with an otherness that never ceases to destabilise us. The experience it provokes is always dual and ambiguous. With its status suspended between *natura naturans* and *natura naturata*, or, as it was called in the Renaissance, *terza natura* (third nature) (Rinaldi 1997), as an intermediate state between wild nature, elusive and instinctual, and cultivated nature, submissive and functional, the garden is defined primarily by its representational function and by its operation as a device that is not only symbolic but speculative: not only does it embody man's relationship with nature, but it also questions it. Whether driven by a nostalgia for paradise lost, for a perfect, primordial unity that is inevitably unrealistic and idealised, or by an equally inevitably unrealistic and idealised project of domination, the garden is founded on a tension between mastery and obsession, between identity and otherness, between order and metamorphosis and between reality and illusion. For in the garden, the relationship with nature is defined by art.

6.2 'Officina Della Meraviglia'

Tu ch'entri qua con mente / parte a parte / e dimmi poi se tante / meraviglie / sien fatte per inganno / o pur per arte » ('You who enter here / to grasp everything / with your mind / will then tell me / if so many wonders / are made by deception / or by art'). Inscription engraved on the statue of the Sphynx, Bosco Sacro de Bomarzo.

It was with the Mannerist Garden that otherness became a programme and a system. From the Bosco Sacro in Bomarzo to Pratolino, from Villa Lante in Bagnaia to Villa d'Este in Tivoli, the garden is the privileged laboratory of Mannerism as anti-humanism (Battisti 1962) – whose historical and current recurrences could be read as the stages of a counter-history of art built on the tension between the human and the non-human. In this counter-history, art is defined in relation to nature, as artifice: an art of devices and technique and an art of concepts, operations and effects, rather than an art of objects. An art that stages a specific, complex and critical relationship between man and nature. Founding both the art of the garden and the more secretive art of the anti-garden, the Mannerist Garden is conceived as a complex and paradoxical device, capable of making the complex relationship between artifice and nature sensitive, spectacular and, one might say, experiential. In contrast to the ordered syntax of classical architecture and the closed structure of the *hortus conclusus* or *locus amoenus*, Mannerism opposes the labyrinthine model of the garden as a 'space of metamorphosis'. As a 'laboratory of the senses' (Praz 1975), a microcosm, the garden is based on an articulated, stratified system in which nature appears transformed by technology: the elements that make it up unfold like the stages of an open-air theatre, where the visitor's wanderings correspond to a symbolic journey across the chessboard of a complex iconology based on the transmutation of the elements and the metaphor of man's path through life as a speculative quest. The alternating alleys, rotundas and maze-like groves, where the vegetation appears unsubdued, but also the fountains and water features, the grottoes, covered in dripping spugne, coral, shells and iridescent mother-of-pearl, where the rituals of cosmic generation are re-enacted (Morel 1998, p. 95; Brunon and Mosser 2014), the theatres of automata invite visitors not only to marvel at the wonders of nature and technology, deployed with great means, but also to lose themselves, as in a forest: It's a journey that doesn't rule out drifting and that doesn't just expose the enigmatic mechanics of metamorphosis, but allows us to put them to the test. Far from being a purely decorative arrangement, the garden is conceived as a device for both scripting and dramatising the experience. Scenarisation first. The Mannerist Garden is a model of what Norman M. Klein (2004) has defined as a 'scripted space', an architecture built around a programme or theme, and composed as a spectacular machine dedicated to producing a specific experience for the spectator. This definition links the Mannerist Garden to the tradition of 'theme parks', which, through the English Pleasure Gardens, the Follies, the Show Gardens, leads to Coney Island's Luna Park, described by Rem Koolhaas in Delirious New York (1994, p. 9) as a model of 'phantom architecture', to Disneyland and the other Amusement Parks deployed by the 'artifice industry' (Eco 2003, pp. 57–58). The spectator's journey is envisaged both as a journey through various locations that make up the park and as a journey through symbols that unfold in a narrative, an initiatory journey. The favole, literary narratives by contemporary authors such as Ariosto, Tasse, Boiardo and Colonna, and ancient mythologies, are the scripts that materialise in the forms of the garden, which are revealed as both a natural environment and an artificial stage.

For the scenarisation of space is matched by its theatricalisation. In the series of attractions and wonders that make it up, the garden functions as an immense theatre without a theatre, where the text is 'embodied' in the landscapes and the characters in the objects, groups of statues and moving automatons. Thanks to technology that reproduces gestures and actions, the park is animated by *accadimenti* (events), whose impact goes beyond that of mere images. Theatre is at the heart of the Mannerist horizon: from Saint Augustine's 'world as a book', we move on to the Theatre of the World. Theatre became not only a metaphor but also the epistemological method par excellence – as demonstrated by the success of the Intermezzi, highly elaborate forms of total theatre that spread throughout the Mannerist period – whose main aim was to dramatise the complex relationship between man and nature. Nature was no longer the stable mirror of a norm, as it had been at the height of the Renaissance, but the 'fascinating and terrifying presence of a variety, plural and dilated, of phenomena [...] an inexhaustible pool of new beings and new forms' (Testa 1991, p. 1), a universe of complex assemblages and metamorphoses. The philosophy of nature that characterises the short Mannerist season does not hesitate to bring to the fore the errors of nature, the bizarre, prodigious, monstrous elements, the anomalies, which become the signs of this new vision of nature, not so much as eccentric departures from its law, but on the contrary as symptoms of its inexplicable and unstable combinatorics. The garden, enclosing living or figured bestiaries, responds to the same encyclopaedic impulse that fills the collections of the studioli, the cabinets of curiosities, and the first museums: synoptic tables of *artificialia* and *naturalia*, theatres for the eyes and the mind, conceptual anamorphoses more than real ones, that only the visitor can reduce to unity, functioning himself as the mirror cylinder that brings together all the dispersed fragments (Lugli 1983, p. 100).

Soliciting the viewer's senses through complex machinery, and the imagination through the evocation of narrative universes, through a technical mastery of the spectacular but also through the systematic use of an aesthetic of excess, dysfunction and deformation, the Mannerist Garden is founded as the 'officina della meraviglia'. Wonder (meraviglia) does not correspond to a state of ordered, limpid and reassuring knowledge, but on the contrary to an affect – a vague and unstable sensation of suspension, bewilderment and simultaneously pleasure – which connects it directly to Edmund Burke's notions of the sublime, Freud's disturbing strangeness, and the surrealist marvellous. At the centre of the Mannerist Garden, a theatre without a theatre is the stage of the automata. Statues of mythical and monstrous creatures, fantastical fauns and modest nymphs, real and fantastic animals, whose movements, powered by complex hydraulic mechanisms, give them the illusion of life – and make the experience both more realistic and more mysterious. After Pratolino, automaton theatres spread throughout Europe: from the Hellbrunn garden in Salzburg, the only one still standing today, to the Neugebaude parks of Maximilian II in Vienna, Rodolf II in Prague and Henri IV in Saint-Germain-en- Laye and Fontainebleau. In the Mannerist Garden system, their function was to showcase the complex relationship between art and the living. As Hervé Brunon (2006) explains, the figure of the automaton needs to be analysed in the light of the debate emerging within the Aristotelian tradition around the mimetic function of art, its ability to simulate or even 'augment' nature, not only in its forms but also in its modes of operation. If, in the framework of Peripatetic theory, movement is the index of life and therefore the living is the model of the automaton, a major epistemological reversal is in the offing, in which the automaton is gradually envisaged as the model of the living (Descartes). It is within this 'animist' horizon of Mannerism that the principle of mimesis is transformed, from representation, an imitation of the forms of nature, to operational simulation, an imitation of the procedures and processes of nature (Falguières 2000, p. 65). This transformation gave rise to a new definition of art, in relation to nature and the living – an ars naturans, the aim of which would be to produce artificial states of nature, experimental natures. Today, at a time when art is confronted with the philosophies of the Anthropocene or speculative realism, which question anthropocentrism and put it to the test of forms of non-human otherness – nature as Gaia (Bruno Latour), or objects – at a time when art seems to want to integrate the living, or ally itself with science and technology, this conceptual project appears to be of burning relevance. By favouring modes of otherness, the use of strangeness, deformation and excess, to stage an 'ontological theatre between the human and the non-human' (Pickering 2009, p. 23), the Mannerist Garden seems to suggest a mirror definition of art as artifice, as anti-nature, as a systematics of the gap, whose aim is to create both pleasure and doubt: an art of crisis, a critical art.

6.3 Black Gardens

« Ogni pensiero vola » ('All thought fades away'.)
Inscription engraved on the door of the Ogre, Bosco Sacro of Bomarzo.

It is the same strategy of otherness that governs the picturesque gardens that spread in the eighteenth century and that Annie Le Brun calls 'black gardens', likening them to the sinister visions of Gothic novels – Bildungsroman in reverse, where the existential journey sinks from light to darkness. After the vertigo of the Baroque and the whims of the Rococo, picturesque gardens can be interpreted as a new exercise in mastering nature and domesticating otherness but also as speculative machines for producing a new form of wonder, the sublime.

But the horizon had changed: at a time when the clarity of the Enlightenment was becoming so blinding that areas of shadow were appearing, when 'a questioning space was beginning to open up between life and its representation', as Annie Le Brun (1982, p. 93) points out, the motive was no longer the esoteric contemplation of original enigmas, but the morose delight of minds exalted by melancholy, faced with a world devoid of any enigma. Just as the Mannerist Garden opposed the clarity of the humanist garden, infusing it with a vibration of restless incredulity, the picturesque garden opposes the rational symmetry of the French garden. The architectural model represented by Le Nôtre was replaced by an ideal that was no longer theatrical, as in Mannerism, but poetic and above all pictorial: as the treatises on the art of gardens that were proliferating at the time explained, true nature could be reached by observing it painted on canvases, as it was not a matter of reason but of suggestion.

In contrast to the French gardens, where the rational scheme was constructed by the symmetries of the terraces and flowerbeds defined by lines of pruned boxwood and expanses of polychrome gravel, the picturesque gardens of the eighteenth century were filled with fabriques, fake ruins, pagodas or oriental temples – a repertoire of otherness and strangeness, for which indexes by type were published – and which constituted the new grammar of the sublime (Mosser 1991; Dixon Hunt 1991). In place of topiary, which bends the insubordination of plants to perfect forms, we prefer to create – as artificial as ever – false wild forests, dramatic waterfalls and even artificial volcanoes. Instead of the overall vision, the endless perspective of squared alleys and pools, which redraw the world as a perfect, luminous geometry, we prefer paths that curve in a contrasting play of copses and meadows, opening onto unexpected landscapes, chasms, dark corners or enigmatic ruins.

From Mannerist theatricality, we move on to an 'aesthetic of atmospheres' (Böhme 2013), paintings arranged according to the gradations of the effects and emotions they evoke.

In a frenzy of excess, some gardens seem haunted by a deep-seated tension, by a heart-rending opposition that transforms them into anti-gardens:

> Two worlds clash phantasmatically: one still apparently assured of its measure, its finesse, its elegance, its clarity – but only apparently –, the other simulating the threat of nature through

> its form of a great nocturnal bird in waiting. Everything that the classical home had excluded from its precincts is suddenly there, infinitely close, infinitely distant. Only the encounter with nature has not taken place: it has been simulated on the chessboard of artifice (Le Brun 1982, p. 105)

The forest is now part of the garden, and the opposition between the inside and outside of the enclosure is just another fantasy: outside the civilised space, outside the advancing city, nature no longer exists, replaced by the landscape, the imprint of the human being stretching across the horizon. So, as Annie Le Brun suggests, at a time when the process of civilisation seems to be triumphant, the darkness no longer hides in the woods, in fear of the non-human, but comes from within, from the vertigo of artifice, as a conscious and nagging illusion, faced with the impossibility of nature, from the doubt that takes root in the cracks of reason, in the fissures of the monumental project of modernity.

It is no coincidence, then, that this same dialectic re-emerges later, when the debate on the postmodern is in full swing. In Collage City (1975), Colin Rowe and Fred Koetter reintroduce the garden not just as a critical model of the modernist city, but as a 'philosophy of life'. They contrast Versailles with the Villa Adriana in Tivoli (whose ruins had so haunted Piranesi), which they see as a Mannerist Garden ante litteram. On the one hand, there is strategy and organic thinking, unity and convergence:

> Versailles, the morality proclaimed to the world; total control, blinding with light [...], the triumph of the modern city, the triumph of generality, the prevalence of the dominant idea, the suppression of the exception" (Rowe and Koetter 1975).

On the other is a thought of disparity and divergence. On the one hand is the rationalist approach of Louis XIV and Colbert and, on the other, an approach that Rowe defines as 'culturalist', based on a complex iconography, a fragmentary collage of 'structural discontinuities and syncopated excitations', not only of objects but also of symbols. And again, on the one hand is a perspective vision, which subjects the journey to the plane, to the vanishing line as a metaphor for linear progress, and on the other is the experience of rambling as free association and vertigo.

Rowe's analysis is important because it underlines another fundamental aspect of the Mannerist anti-garden: more than its synchronism as a mnemonic machine (Mosser and Nys 1995), its anachronism – in the dense present of experience, the stratifications of memory accumulate, all the times of the world blend together. In essence, what the garden teaches us is that time is nothing but metamorphosis.

6.4 Theory of Displacement

So, if we reintroduce the dimension of time into the experience of the garden and if we give priority to the dimension of wandering and the journey, we might wonder whether the origin of the garden might not be the path, rather than the clearing: another form of breakthrough in the blackness of the forest, implying not a stop, but a crossing: more than a location, a displacement. This is what Giuseppe Penone seems to be suggesting, in an annotation accompanying the creation of the *Giardino delle sculture fluide* (Parco della Venaria Reale, Turin, 2003–2007):

> The garden begins when a man walks on the ground, and advances into the space of the plant, of the mineral. His actions are imprinted on the ground, and the minute realities that his footsteps encounter memorise his presence. The shrubs moved by a force other than the wind, the broken twigs, the bent and trampled grass, the tiny invisible animal life disturbed by the footsteps, all bear witness to the passage, and are reminders of man's journey. The perception of these innumerable small events, the reflection, the observation, the amazement that accompanies the path of one who advances with his gaze lowered towards the earth, and his suspended thought, impregnated with the sky, harmonise the senses... from that moment, the memory of his presence becomes fixed in the place. The systematic organisation of this memory, its structuring, the desire to ritualise the route and its repetition generate the garden (Penone 2015, np).

And when the path twists in on itself, the line becomes a spiral, the repetition becomes an obsession, and the darkness folds in, offending the sky, the path becomes a labyrinth. The labyrinth is not just a syntagm in the syntax of the garden, but one element among others in the park and the path, as at Villa d'Este or Pratolino, as a paradigm (Brunon 2008) and as a founding principle. By revealing the garden's maze-like nature, it returns architecture to the archetype of wandering as the foundation of the quest for self and seems to suggest, through its sinuous volutes, its lures and its endless dead-ends that it is not the exit that counts, but the traversing: to traverse a garden is to be traversed by the garden, to no longer be the same.

References

Battisti E (1962) Antirinascimento. Feltrinelli, Milano
Böhme G (2013) Atmosphäre. Essays zur neuen Ästhetik. Suhrkamp
Brunon H (2006) Ars naturans, ou l'immanence du principe : l'automate et la poétique de l'illusion au XVIème siècle. In: Pieper J (ed) Das Château de Maulnes und der Manierismus in Frankreich. Beitrage des Symposions am Lehrstühl für Baugeschichte und Denkmalpflege der RWTH Achen 2001. Aachener Bibliothek, 5. Deutscher Kunstverlag, pp 275–304
Brunon H (ed) (2008) Le Jardin comme labyrinthe du monde. Métamorphoses d'un imaginaire de la Renaissance à nos jours. Éditions du Louvre-Presses Universitaires Paris-Sorbonne
Brunon H, Mosser M (2014) L'imaginaire des grottes dans les jardins européens. Hazan
Dixon HJ (1991) Gardens and the picturesque: studies in the history of landscape architecture. MIT Press
Eco U (2003) Dalla periferia dell'impero. Cronache da un nuovo medioevo, Bompiani
Falguières P (2000) Mécaniques de la marche. Pour une pathétique des images animées. In: Un siècle d'arpenteurs, Les figures de la marche, exhibition catalogue. Musée Picasso. RNM, Antibes
Harrison R (1992) Forêts, Essai sur l'imaginaire occidental, Flammarion
Klein NM (2004) The Vatican to Vegas, a history of special effects. New Press
Koolhaas R (1994) Delirious New York. Monacelli, New York
Le Brun A (1982) Les Châteaux de la subversion. Gallimard
Lugli A (1983) Naturalia et Mirabilia. Il collezionismo enciclopedico nelle Wunderkammer d'Europa. Mazzotta

Morel P (1998) Les grottes maniéristes en Italie au XVI siècle. Macula
Mosser M (1991) Les architectures paradoxales ou petit traité des fabriques. In: Mosser M, Teyssot G (eds) Histoire des Jardins de la Renaissance à nos jours. Flammarion, pp 259–276
Mosser M, Nys P (eds) (1995) Le Jardin, art et lieu de mémoire. Les éditions de l'imprimeur
Penone G (2015) Giardino delle sculture fluide. leaflet, Torino-Parco della Venaria Reale
Pickering A (2009) The cybernetic brain: sketches of another future. Chicago University Press, Chicago
Praz M (1975) Il giardino dei sensi. Studi sul manierismo e il barocco. Mondadori
Rinaldi A (1997) La ricerca della "terza natura": artificialia e naturalia nel giardino toscano del Cinquecento. In: Fagiolo M (ed) Natura e artificio. L'ordine rustico, le fontane, gli automi nella cultura del Manierismo europeo. Officina Edizioni
Rowe C, Koetter F (1975) Collage city. The Architectural Review CLVIII:942
Testa F (1991) Spazio e allegoria nel giardino manierista. La Nuova Italia

Part II
Interdisciplinarity and Multiscalarity of the Urban Green System in the Face of Contemporary Environmental, Cultural and Social Emergencies. Theories and International Case Studies

Part II
Interdisciplinarity and Multiscalarity of the Urban Green System in the Face of Contemporary Environmental, Cultural and Social Emergencies. Theories and International Case Studies

Chapter 7
About the Garden: Which Figures and Which Reinterpretations Provide Answers to Contemporary Questions Without Betraying Its Meaning

Alessandra De Cesaris

Abstract The chapter questions the possible role of the garden – a place built with predominantly natural elements – in the contemporary era characterised by environmental, cultural and social emergencies. The history of the garden – an archetype of our culture – reflects that of every civilisation, and the garden represents a link between nature and architecture. In the past, each era has given its own particular interpretation of the garden, providing its own responses to the specific needs of society. Today, faced with the demands of the contemporary world, new reinterpretations and new layouts seek to go beyond the traditional notion of the garden, an archetype of our culture, without betraying its profound meaning. It is therefore essential to combine quantitative and qualitative data, integrating the current demands for more green spaces in the urban fabric of European cities with sustainable solutions also from a formal and aesthetic point of view.

Keywords Garden · Green · Park

7.1 Introduction

In recent years, we have witnessed an increasing focus on the design of green spaces, attentions aimed at identifying ways for contemporary man to relate to nature in a way that is conscious and representative of his own era. We are also questioning the role that the natural component can play in the regeneration of our cities.

The ecological instance to green up our cities, to improve air quality, identifies large-scale infrastructures, and there is talk of green infrastructure, of urban forestation, and few references to the entity that is the garden as if it were irretrievably linked to times gone by, as if it had ceased to communicate innovation and as if the

A. De Cesaris (✉)
Department of Planning, Design, and Technology of Architecture, Sapienza University of Rome, Rome, Italy
e-mail: alessandra.decesaris@uniroma1.it

F. Dal Falco (ed.), *Beyond the Garden*, Designing Environments,
https://doi.org/10.1007/978-3-031-50567-6_7

garden were the bearer of a relationship between man and nature of bygone days. What form to give to the garden of the ecological era asks Gilles Clément in wishing for the disappearance of its enclosure (Clément 2012).

In this tense, quest dictated by the ecological need to repair the damage done by man to our planet and to improve the quality of our built world – on the whole, with very few exceptions, built without any quality at all – the meaning and figurations of the terms green, garden, park, and landscape overlap and contaminate each other, blurring into each other.

It is possible that from these contaminations arise solutions and figurations capable of identifying new ways for man to configure – reinforcing – his relationship with nature. It is a relationship dense with criticality within a highly artificial environment, made of stone, asphalt, cement, but also of sheet metal, eternit, mud and straw.

Is it possible, who knows?

In order to better understand the possible exits, it seems appropriate to me to make some clarifications on the origin, meaning and role of the figures of garden, park and green, clarifications aimed at understanding how it may be possible today to go beyond the garden without betraying its deepest sense.

Because the garden may not be green, the green may be not only red (this is the case of the wild prunus for example) but also multicolored, and the park is not a garden, even though it may contain it within it; this is the case, for example, of the English garden within the park of the Royal Palace of Caserta wanted by Maria Carolina, wife of Ferdinand IV of Bourbon, where the criteria that preside over the respective compositions blur into one another. We are in fact faced with a park designed by Vanvitelli as an extra-large garden and a garden designed as what would become parks. But much of the fascination of this place consists precisely in crossing boundaries. Boundaries, contaminations, hybridisations are therefore to all intents and purposes design elements.

7.2 Garden

The garden – a space built with tree essences – is an enclosed place intended for walking and recreation. It has agricultural origins of a utilitarian nature; the grammatical elements that make up the garden – the alignment of plants, the design of irrigation canals, the size of the compartments – were derived from agricultural needs, and then the productive reasons were associated with aesthetic reasons; 'aesthetic distillate of agricultural civilisation' defines the garden (Petruccioli 1995, p. 85).

There is no civilisation in the world that has not felt the need to have its gardens. The garden as a necessity is therefore as man's primary need (Grimal 1974).

It is from this primary need, from the necessity of the garden, that the magnificent object that is the Persian carpet in its garden variant originates. A mat of millions of interwoven knots that reproduces the geometric structure of the archetypal four-part *chahar bagh* (literally four gardens). A geometrically patterned border defines and delimits the space, protecting it from hostile surroundings, while inside, a series of bands with zigzagging motifs recall the flow of water and subdivide the fields adorned with geometrically patterned nature in the form of trees, bushes, flowers, birds and fish. It is said that during the long, cold winters, the Persian king Khosrow had a magnificent carpet spread out on the floor of his palace at Ctesiphon, called the Spring of Khosrow to recall the spring season: 'the border was embroidered with emeralds, the centre depicted avenues, streams, a garden, an orchard and a wheat field. Contemplating the designs, Khosrow erased the snow and the cold, and the joys of spring seemed closer and more familiar to him' (Citati 1977, p. 75).

There have been many evolutions of the garden over time and around the world, and each culture has elaborated different garden figures over time and attributed different symbolic values, and common to all civilisations however is the reference to an earthly paradise.

The Islamic Garden – which has spread over a vast geographical area, from Spain to India via the Maghreb – has its origins in the Persian archetype of the *chahar bagh*; it is a space enclosed by buildings or walls subdivided into four parts by two axes in which a third zenith axis – which brings the cosmos and the tangible world into harmony – is always implied (Petruccioli 1995).

In the geographic area of the Fertile Crescent and the Mediterranean, the garden has played a decisive role in the process of anthropisation; in fact, the house is built around the garden, and this empty space becomes the ordering element of the urban fabric (Petruccioli 2021). It is an enclosed, protected space, which although invisible to most has played and continues to play a fundamental role in the daily life of those who live there.

In the West, in our historical cities, this decisive role of the garden void in the formation of the urban fabric is missing, and the garden and nature more generally take on other values, other, less structural meanings; certainly the nineteenth century city extracts from the garden the elements to build its decorum, the tree-lined avenues of parks cross the borders, and define the new urban expansions, but the garden at our latitudes does not take on the form of an ordering element of the urban fabric.

From a purely spatial point of view in Europe, the *hortus conclusus* of the medieval garden – enclosed by high walls of castles and monasteries – was succeeded by the Renaissance Garden, the private delightful place of noble villas. Here, the design became strictly geometric; nature was something to be controlled, to be dominated according to the use of geometric and perspective rules that organised the composition of the whole. These were also characteristics of the later Baroque Garden and would instead disappear in the landscape gardens of the late seventeenth century first half of the nineteenth century. The excessively formal and geometrical design gives way to a freer composition that tends to emphasise the beauty of wild nature. The essentially static perspective view set on an axis is replaced by a series of winding paths that stage different views, the attempt being to recall the natural environment as much as possible. The landscape garden thus stages the idea of nature that the culture of the time was elaborating.

Gardens of various forms and representations respond to various ways of understanding the relationship between man and nature, characterised however by one constant: the presence of an enclosure (Figs. 7.1, 7.2, 7.3, and 7.4).

The garden is in fact an enclosed, protected place, and the concept of enclosure is inherent in the very etymology of the name Garden: the term in fact derives from the

Fig. 7.1 Enclosures: Chahar bagh Carpet, early eighteenth century, Tehran Carpet museum. (Source: Author's photo)

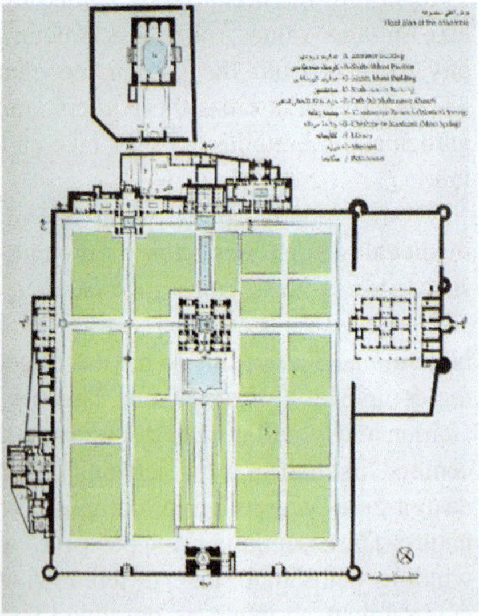

Fig. 7.2 Enclosures: Isfahan, Planografia SedisRegiae, Engelbert Kaempfer, Kashan, Bagh e Fin. (Source: Author's elaboration)

Fig. 7.3 Enclosures: stonehenge. (Source: Flickr, Giorgio Raffaelli)

Fig. 7.4 Enclosures: Parc Citroen, Gilles Clement and Primary School in Fontana Candida, Labics Studio. (Source: Author's elaboration, from http://www.gillesclement.com/cat-banqueimages-andre-tit-banqueimages-andre and https://www.labics.it/project/92)

Indo-Germanic root Gart or Hart, which has the meaning of 'to enclose, to surround' (Calzolari 1968, p. 465).

7.3 Park/Urban Park

According to the Dictionary of Architecture and Town Planning, the origin of the word derives from the Latin parcus, which in the fourteenth century turned into barco (game enclosure) and differs from the garden in the more natural appearance of the green arrangements and the larger size.

In many seventeenth and eighteenth century villas in France and Italy, the park was not yet a separate entity but an intermediate element between the garden and the natural surroundings.

Although some gardens, in particular the gardens of royal villas (the park of the Reggia di Venaria, the park of the Reggia di Caserta, the park of the Reggia di Portici) are defined as parks mainly in relation to their vast size, the park as we understand it came into being in the mid-nineteenth century.

From this date onward, hunting reserves and gardens belonging to aristocratic villas break through their enclosure walls and welcome, at least in some of their parts, the general public. The park thus originates from the matrix of the garden but expands its dimension to relate to the scale of the nineteenth century city and becomes democratised, from a private place for the use and consumption of a specific social class it becomes a public space. In the past, access to the gardens and parks of villas was in fact only granted on certain special occasions. The poor classes had other ways of relating to nature, poor gardens, for example, as we shall see later.

Furthermore, in the peripheral areas of cities, patches of countryside are transformed into parks, spaces for environmental and social rebalancing. In this way, a well-defined typology is born, the urban park, a typology made mainly of natural materials that contrasts with the stone city, a place for outdoor recreation and breathing healthy air. The enjoyment of nature, until then reserved for the aristocratic classes, thus became the patrimony of the working classes. It is a new way of relating to nature that has a strong hygienic and sanitary value and provides answers to the phenomena of urbanisation of the poorer classes, a new way in which the purely aesthetic aspect loses importance over the functional aspect.

But nineteenth century culture also defined another way of relating to nature: the garden city. Again, the garden or a garden city, to decongest large cities by decentralising the population into satellite towns surrounded by greenery on public land. This is Howard's utopia, from which two garden cities took shape, where one could live in harmony with nature and lead a healthy life and annul the dichotomy city-countryside. It should be emphasised that in the garden city envisaged by Howard, the non-trivial question of land ownership remained a fundamental issue because his ideal city – a city in the garden – built on public land was a far cry from that juxtaposition of houses with gardens – from that city made up of private gardens that has subsequently urbanised the suburbs of half the world with enormous and unsustainable land consumption.

7.4 Green

The so-called 'green' represents the way in which the modern interprets the relationship with the natural element. It is an abstract term, very abstract. In fact, it does not contemplate nuances, and we know that, in nature, green has infinite nuances; it also does not contemplate the passing of time and the seasons, and in fact it does not contemplate the colours of flowering nor the less garish colours of autumn. Nor does it contemplate morphological information. What shapes what figures define this abstract green? Nature in the modern enters the city as a continuous carpet, as an identifiable surface through its square meters, and it thus becomes a standard and in becoming a quantity excludes any formal reference.

In Italy, law 1444 of 2 April 1968 established the minimum and mandatory allocation of 9 square meters of public green space per inhabitant; it is a law that attempted to put an end to the damage caused by the cementification of the territory and building speculation, although in reality, it often defined unresolved and desolate spaces, SLOAP Space Left Over After Planning. The term was casually coined by Leslie Ginsburg in the course of a discussion at the Birmingham School of Planning to indicate a plan in which vast portions of unused empty areas dominated. Since then, the acronym SLOAP indicates those spaces – perhaps green spaces – resulting from the arrangement on the ground of the fullness of buildings for which no design thought has ever been developed (Ginsburg 1973).

Of demotion or downgrading of the garden to a green area Rosario Assunto speaks explicitly when he writes 'that architectural-urban functionalism instead of the words gardens and parks adopts words of pure and simple communication (indicating extensions and not quality) such as points and areas or zones or even space' (Assunto 1982, p. 28).

In recent times, the awareness of the fragile state of our ecosystem caused by global warming, climate change and the increase in carbon dioxide emissions has progressively overshadowed the formal, aesthetic aspect of green design, and the role that urban vegetation can play from an ecological point of view is increasingly emphasised. Urban greenery is thus defined 'as green areas within which nature can manifest itself with varying degrees of freedom, with or without the presence of man. They are areas characterised by the presence of soil and spontaneous and non-spontaneous vegetation, regardless of their dimensional characteristics, urban territorial or building scale of reference' (Lassini et al. 2014, p. 50).

Today, the term green includes other meanings, meanings related to the ecological need to take care of one's environment. *Green* has thus become not only the attitude of the individual citizen in respecting his or her environment but also a political party that carries on battles for a precisely green transition to the city of the future.

And today through what strategies, methods and figurations can we reinterpret the relationship between man and nature and what role can nature play in the regeneration of the contemporary city?

How can parks, greenery and gardens contribute to the regeneration of our cities? How can the notion of standards become the bearer of innovative values? Is there still room for the survival of the garden? Can the garden still represent an archetype to be reinvented? Can the garden with all its enclosure become the bearer of new values? If so, how? Utility and Education writes Joseph Rykwert (1981).

7.5 The Garden: Evolutions/Interpretations/ Contaminations

Over the centuries, there have been many reinterpretations of the garden. In the early nineteenth century, botanical gardens (an evolution of the botanical gardens created in the mid-sixteenth century) and zoological gardens (an evolution of the menageries present since antiquity in private parks and noble villas) were created. Within these gardens with a didactic and scientific purpose, there are acclimatisation greenhouses for exotic plants, often of great beauty; zoos, now transformed into bioparks, house extravagant artefacts designed in the style of the country of origin of the beasts. Overall, they collect specimens, sort, catalogue and disseminate.

Extraordinary gardens are fantastic places within the urban fabric, which combine leisure and entertainment with an educational value and an important scientific function.

7.5.1 Above the Ground

In the twentieth century, Le Corbusier, mindful of the lessons of Hennebique and Hènard, codified the roof-garden. The garden is detached from the ground, in some cases flying over the roof and becoming a ceiling, a coverage.

It was in fact Hennebique, the inventor of reinforced concrete, who experimented with the potential of this material in his villa in Bourg-la Reine (1902–1904). The roofing slab of the villa was covered with a meter of earth to create an aerial garden widely publicised by Hennebique to promote the Systéme Hennebique he patented in 1892.

Eugène Hénard in the same years also pursued the dream of a Paris whose roofs were transformed into gardens, but it would be Le Corbusier with his 5 points and his realisations who would codify a new typology: that of the roof garden.

'Instead of tiles or slate, and thanks to reinforced concrete, marvelous walks... The whole city, the whole of Paris can be crowned with enchanting places. Instead of tiles or slate, roof gardens. A show dedicated to the devotion of Paris'; this is the caption of a reproduction of the extraordinary roof-garden of the Beistegui penthouse (Croset 1981, p. 27).

Le Corbusier's garden detaches itself from the ground in various ways, defining worlds with strong aesthetic qualities. In the Beistegui attic, the end of the promenade is configured as an open-air room of surrealist memory where four white walls enclose a grass floor and alienate iconic elements of the Parisian landscape from their context.

In the Ville Savoye, we find an open-air room, perhaps a reminder of the patios of Mediterranean memory. It is a room from which to view distant portions of the landscape that through the device of the ribbon window are, once again, estranged from their context and captured within the geometric grid of the house.

'The Roof Garden: Technical Reason and Aesthetic Ideal' is the title of an interesting article by Pierre Alain Croset in times not spoiled by ecological excess emphasises the technical advantages of the roof garden.

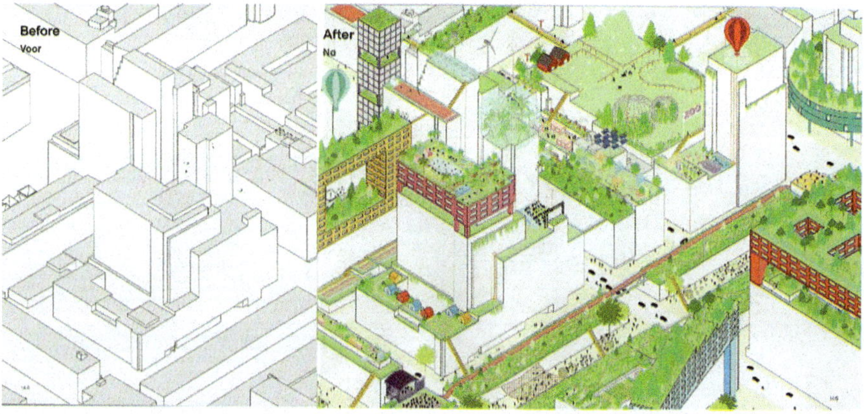

Fig. 7.5 Rooftop MVRDV. (Source: MVRDV)

It is a strategy that is today supported by many on the basis of arguments relating to energy efficiency, environmental sustainability and biodiversity. However, nothing detracts from the possibility of defining spaces that are also pleasant for humans, and the pandemic has taught us the potential value of these surfaces, especially in the Mediterranean area and in the Islamic city area (Fig. 7.5).

A potential immediately understood and developed by MVRDV in the Rooftop Catalogue that 'shows how roof use can be a solution for the scarcity of space in the city and how it can help to prevent the city from expanding outwards into rural areas' (Fig. 7.6). On the top of densifying cities, there is plenty of room to fix the housing challenge, energy transition, climate adaptation and inclusiveness. In this combination of challenges, one helps the other. Among the many hypotheses we find are: adding facilities for the block, adding a new type of housing for a more inclusive neighbourhood and last but not least 'Adding Gardens (often short in the city centre)' (MVRDV 2021).

The catalogue has been commissioned to the architects by the Rotterdam municipality in order to find select propose innovative ideas to make use of the city's roofs for Rotterdam's 18.5 km^2 of – mostly empty – flat roofs.

But it is perhaps in the gardens of the Immeubles Villas, in those double-height voids that pierce the compactness of the building and introduce porosity into the thickness of the building block, that many possibilities for the regeneration of our cities are inherent today.

The domestic space of the immeubles villas is organised precisely around the full-height void of the roof garden, which differs from those small balconies decorated with greenery where it is impossible to stand 'around a small table with four chairs': this, according to Hertzberger, is the minimum dimension of an open-air space. Roof gardens that bring light into the home and provide it with a liveable, usable outdoor space.

This theme was recently investigated by me in a series of case studies on the Tehran metropolis, research conducted as part of the Cultural Agreements I signed between Sapienza and Iranian universities (Fig. 7.6).

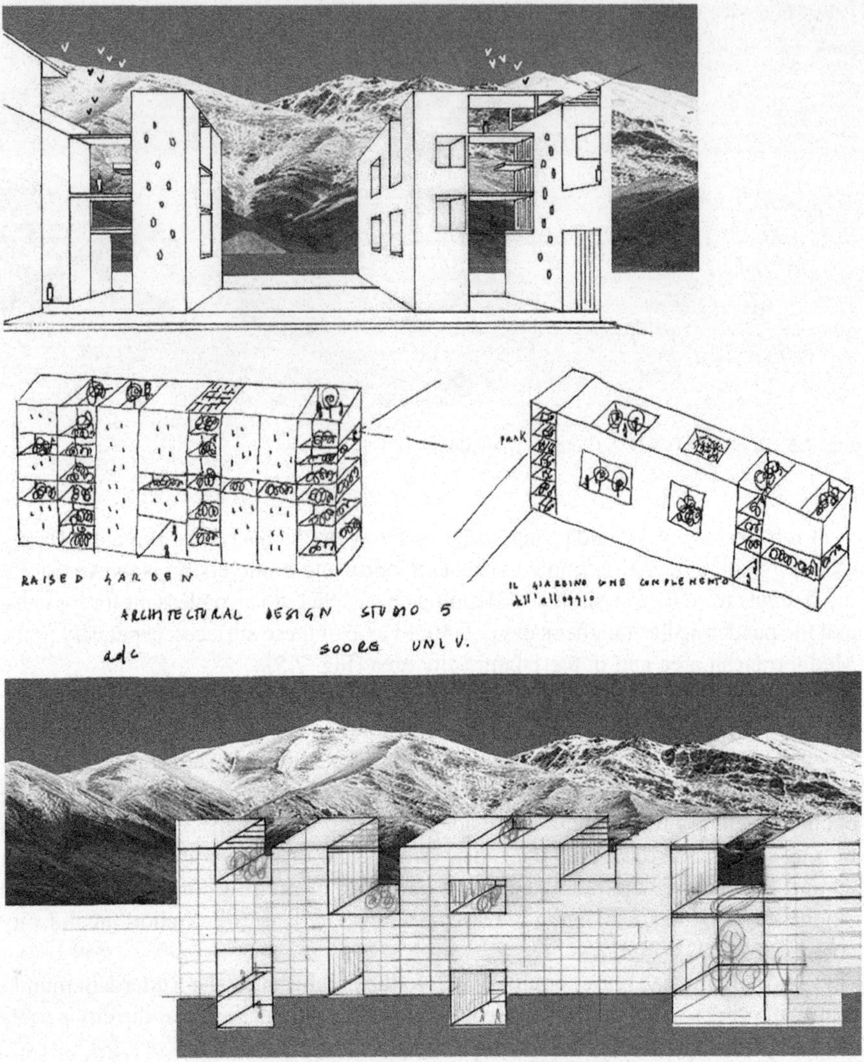

Fig. 7.6 Tehran, raised garden buildings. (Source: Author's elaboration)

We experimented with ways of combining spatial quality and 'porosity' in the dense built fabric of Tehran metropolis of 14 million inhabitants, which is progressively verticalising and erasing all forms of garden, a fundamental element of the original *forma urbis* of the city and culture of this country. Tehran was indeed a city of gardens (De Cesaris 2020).

We have proposed buildings that introduce porosity into the dense built texture, raised garden buildings, which re-invent, re-interpreting at high altitude, that splendid, wonder that is the Persian garden. We investigated the possible distributions of these green voids (volumes and not surfaces!) in relation to ventilation, cooling

(Tehran is a torrid city in summer) and the social use of these, in a city decidedly lacking in public spaces. We have therefore attempted to compensate the city for the greenery and gardens that have been progressively expelled not only from domestic life but also from the urban fabric.

7.5.2 At Zero Level

These are the many interesting potentials of green raised from the ground; in the context, therefore, of a possible redefinition of building types, but at zero height, how can we act today? What possible renewed relationship between man and nature can we glimpse? What forms and what role can the garden and its many variations/geminations play in the redevelopment of our cities?

Perhaps, it is precisely in the tiny dimension of the garden – a dimension that allows it to be inserted with relative ease into the mesh of the urban fabric – that new public spaces of proximity can take shape, unless, as can be glimpsed in the visions of some, we wish for the dissolution of the city in nature.

An increasingly popular way of relating to nature with informal, inexpensive interventions is the creation of ephemeral gardens and urban vegetable gardens in marginal, degraded, unused areas. It is a way of regenerating the urban fabric and creating socialising spaces with not only a social but also a therapeutic and productive value.

In reality, urban gardens are a typology with distant roots. 'During the Napoleonic wars (1803–1815), a debate arose in England on whether or not it was expedient to grant land to workers so that they could cultivate fruit and vegetables in their free time The starting point of this controversy, which led to the creation of the first Poor Gardens, will be the enactment in 1819 of a law (Select Vesties Act) granting parish priests and overseers of anti-poverty laws the authority to lease plots to the unemployed. From there onwards the talk of allotments/(allotments) began, which would become an increasingly urban phenomenon during the 19th century' (Beruete 2016, p. 353).

From the second half of the nineteenth century to the present, 'these places of community agricultural production have shown great resilience by changing character and even users, in relation to the social changes that have affected the populations of the city and the transformations of urban landscapes' (Panzini 2021, p. 9). So writes the author concluding with the conviction that urban gardens, in the renewed forms they will take, are destined to play an important role in the configuration of urban settlements in the coming decades.

Variations on the theme are represented by the Guerrilla Gardens, spaces of resistance and social protest where the desire to build and care for a garden is combined with political activism.

'The Green Guerrilla, born in Manhattan in the early 1970s, was the first non-profit association to claim community gardens as a political tool to serve the regeneration of degraded urban areas and a means to promote the involvement of neighbourhood communities in solving their problems' (Beruete 2016, p. 13).

The Lower East Side, Loisada community gardens are gardens as big as blocks or as small as flower beds and arise from the spontaneous initiative of individual groups of citizens to revive abandoned and degraded areas of Manhattan. Experimental, non-codified solutions of great creativity and expressiveness.

Franco La Cecla, in the introduction to the book that tells the story of these gardens, warns of the danger of leaving the poor and marginalised the right to make do, the danger of a system that gets rid of welfare and puts all the costs of 'making society' on citizens, on their informal making. It is a kind of right to creative unemployment (Pasquali 2008).

The issue of citizens' creativity re-appropriating spaces and the logic of do-it-yourself and freedom of expression, which is often combined with the inertia of the bodies in charge of caring for common spaces, is a matter for reflection and consideration. Actions dictated by free creativity should not interfere with the quality and design of the common space. What would happen in Unter den Linden if everyone planted an essence other than a lime tree in memory of an event or a person who had passed away? This is increasingly common in the boulevards and streets of our cities.

And then there is water, a garden is born and grows in the presence of water, which may or may not be spectacular but is a crucial element. And today the water crisis of this resource directs us toward what form of gardens? We can envisage gardens with essences that require little water, but also gardens that collect rainwater or phytopurification gardens. Gardens that therefore combine aesthetic requirements and environmental values.

> In the workshop H2O Multilevel rivers. Climate change and urban spaces. Multilevel rivers for contemporary challenges" organised in September 2022 by myself with my colleagues Andrea Iacomoni and Chiara Ravagnan we hypothesised for the city of Isfahan a system of phyto-depuration gardens in empty plots or plots occupied by dilapidated buildings in the Safavid period building fabric.

Together with the students, we hypothesised and designed a system of gardens that collect white water waste from the surrounding fabric (see Figs. 7.7 and 7.8).

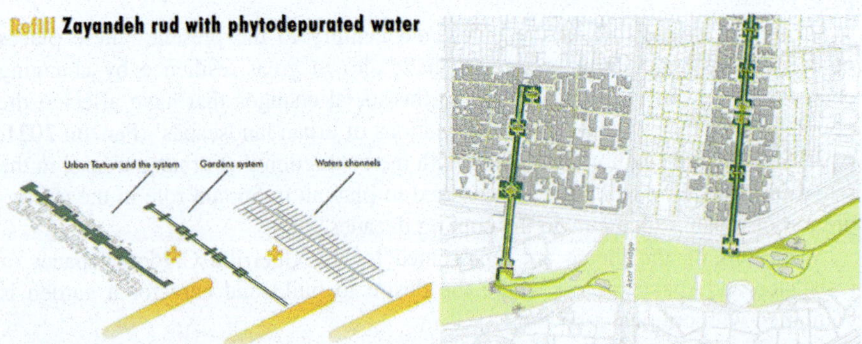

Fig. 7.7 Isfahan, phyto-purification gardens. Workshop H_2O multilevel rivers. (Source: Author's elaboration)

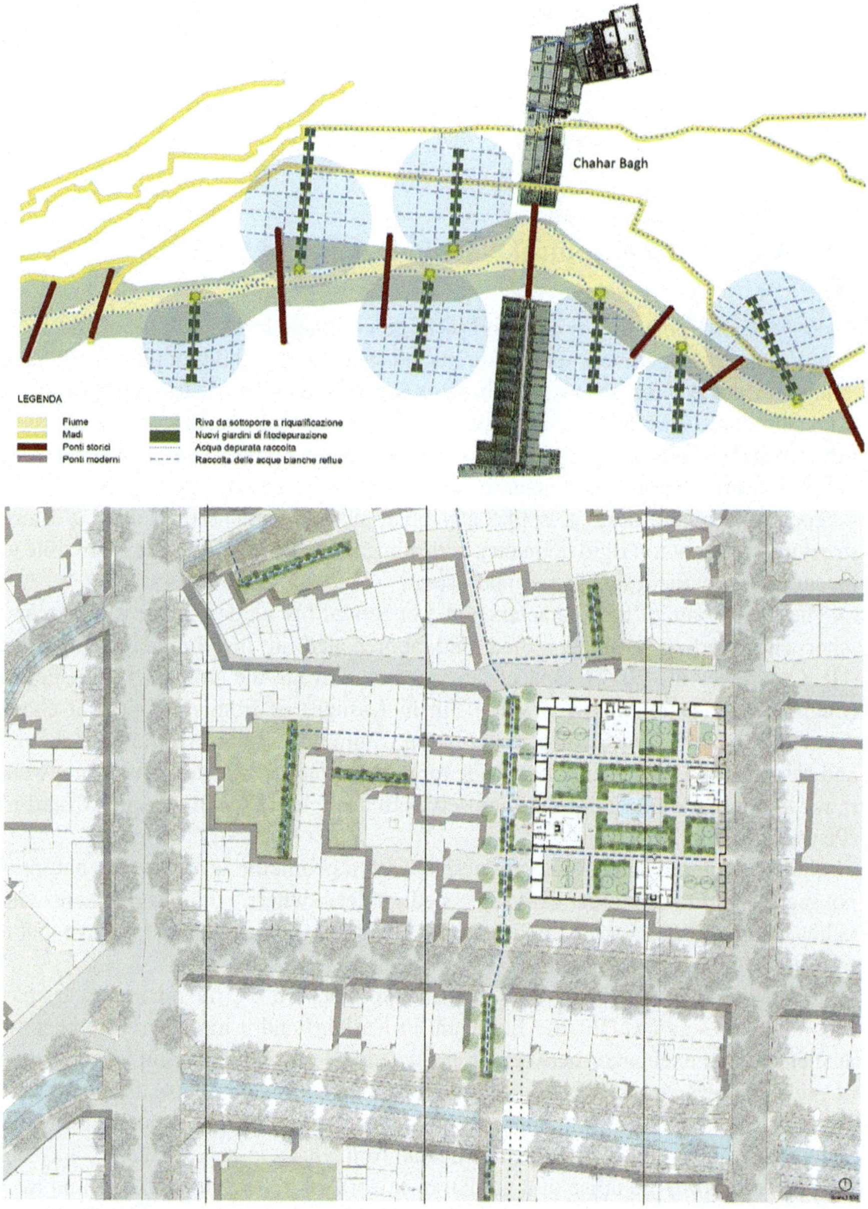

Fig. 7.8 Isfahan, phyto-purification gardens, Sara Azzam thesis degree. (Source: Author's elaboration)

Gardens reinvent the archetypal Persian garden in a contemporary key, gardens enclosed by walls with small urban services and in the centre phyto-purification tanks; from here, the purified water is channelled, depending on the slope of the land, to *Madi* or to the river Zayanderud today dry. This would define corridors of urban regeneration on the right and left banks of the *Zayanderud*, within a dense urban fabric currently lacking in greenery and would create garden systems parallel to the historical urban axis of the *Chahar bagh*, garden systems mindful of the structural relationship between garden and city that characterised Isfahan's urban form.

7.6 An Autobiographical Note in Defence of Gardens

I was born in Rome in a house with a large communal garden, one of the most beautiful in Prati neighbourhood. A protected space where children played and play freely and the elderly observe the blossoming of plants, smell the scent of citrus fruits and come into direct contact with the rhythm of the seasons. It is a space of freedom and socialisation within the condition of contemporary solitude and an oasis of peace from the noise of Roman traffic.

Today, as in the past, it is like any other place, usable for all ages, a space structured by a not too rigid geometry with some secret places, where it is possible to invent infinite varieties of games and uses.

These are some of the characters and wonders of my garden, other wonders and other characters I have identified and appreciated in my Iranian wanderings (De Cesaris et al. 2017). Spaces are protected from the aridity of the desert (Bagh-e Shazdeh in Mahan, Bagh-e Fin in Kashan) or from metropolitan chaos (Tehran), authentic places of contemplation (Shiraz). Places of meditation, refined and passive, the Persian garden is made for contemplation: 'the Persians do not walk in the garden as we do but limit themselves to a single perspective writes Chardin' (Petruccioli 1995, p. 11).

Gardens that are sometimes productive, with a geometric layout where nature is not regimented to any great extent. Fenced gardens, where – in disagreement with Gilles Clément's call for the abolition of the fence – (Clement 2012) I believe that it remains a fundamental element.

It may not be a wall, it may become porous, it may be inhabited, it may accommodate services (Bagh-e Fin again in Kashan) and it may therefore become an element that exchanges between an inside and an outside, a margin where a zone of exchange between different realities is activated, but it cannot lose its character of delimitation, of passage from one place to another, from urban to vegetal, from wild vegetal to architecturally landscaped vegetal, from chaos to quiet, but also, why not the opposite, from the noise of a garden in celebration to the quietness around it.

The awareness of the fragility of ecosystems calls us all to new responsibilities, and as Gilles Clement argues, the emergence of ecology leads us to consider the existence of one great enclosure that is the earth (the planetary garden) (Clement 2011). In this great enclosure, we'll welcome differences, we'll 'do as little as possible against' (Clement 2011), we'll demolish asphalt and concrete, we'll give space to mixtures and hybrids, but we also have to contemplate the presence of other micro-enclosures capable of preserving that precious and fragile entity that is the garden.

References

Assunto R (1982) I giardini della parola e la parola dei Giardini. Artibus et Historiae 3(5)
Beruete S (2016) Giardinosofia. Una storia filosofica del giardino. Ponte alle grazie
Calzolari V (1968) Voce Giardino. In: Portoghesi P (ed) Dizionario Enciclopedico di Architettura e Urbanistica. Romanoeditoriale
Citati P (1977) La primavera di Cosroe. Rizzoli
Clément G (2011) Il giardino in movimento. Quodlibet
Clément G (2012) Breve storia del giardino. Quodlibet
Croset PA (1981) Il tetto-giardino: ragione tecnica e ideale estetico. Rassegna 8
De Cesaris A (2020) Case iraniane. Il valore dl vuoto. FrancoAngeli
De Cesaris A, Di Giorgio G, Ferretti L (2017) Attraverso l'Iran. Città architetture paesaggi. Manfredi edizioni
Ginsburg L (1973) Architectural Review 920, October 1973
Grimal P (1974–2005) L'arte dei giardini. Una breve storia. Donzelli
Lassini P, Sala G, Bertin L (2014) Spazi verdi: manuale di progettazione e gestione agro-ambientale. Edagricole
MVRDV (2021) Rooftop catalogue. Rotterdamse Dakendagen
Panzini F (2021) Coltivare la città. Storia sociale degli orti urbani nel XX secolo. Derive Approdi
Pasquali M (2008) I giardini di Manhattan. Storie di guerrilla gardens. Bollati Boringhieri
Petruccioli A (1995) Il giardino come anticipazione della città. Storie parallele. In: Petruccioli A (ed) Il giardino islamico, architettura, natura, paesaggio. Electa, p 85
Petruccioli A (2021) Il giardino islamico tra storia e architettura. Unità nella diversità. L'industria delle costruzioni 479
Rykwert J (1981) Il giardino del Futuro fra estetica e tecnologia. Rassegna 8

Chapter 8
Green Roofs as a Mainstreamed Nature-Based Solution Tackling the Challenge of Biodiversity Loss

Chiara Catalano

Abstract Green roofs (GRs) belong to the third type of NbS, namely, the creation of new ecosystems synthetically defined as surfaces detached from the ground, spontaneously colonised by plants, or intentionally greened. Their strength lies in the multiple benefits (co-benefits) they offer to single buildings and the urban environment as a whole: from the absorption of air pollutants and the reduction of energy consumption in buildings to the provision of biodiversity. Concerning the latter service, if GRs are designed according to the principle of restoration ecology following the habitat template approach, they can play a key role as stepping stones to becoming part of the urban ecological network and the urban green infrastructure. Conceiving GRs as landscapes instead of flat homogeneous surfaces will improve biodiversity, for example, by varying substrate type and thickness, adding small temporary ponds and foreseeing areas with scattered vegetation. Among others, a way to select plant species and communities for green roofs is to take the phytosociological classification as a template given by the characteristic, diagnostic and recurrent species of natural stands. An accurate preliminary site analysis is essential to select the proper natural template and to replicate the edaphic conditions characterising it.

Keywords Biodiversity · Landscape architecture · Green building · Green infrastructure

---

C. Catalano (✉)
National Research Council (CNR), Research Institute on Terrestrial Ecosystems (IRET), Porano, Italy

National Biodiversity Future Center (NBFC), Palermo, Italy
e-mail: chiara.catalano@cnr.it

F. Dal Falco (ed.), *Beyond the Garden*, Designing Environments,
https://doi.org/10.1007/978-3-031-50567-6_8

8.1 Introduction

Among the aims of the European Green Deal (EGD), the 'no-net emission of greenhouse gasses', the decoupling of economic growth from resource use and the inclusion of people and places are pivotal (Fetting 2020). In 2021, within the roadmap towards a climate-neutral Europe, the New European Bauhaus (NEB) meant to connect the EGD 'to our living spaces and experiences' through initiatives, projects, and ideas, which are 'enriching', 'sustainable' and 'inclusive'. A core part of the EGD is the EU Biodiversity Strategy for 2030, which accounts also for the design and maintenance of urban green spaces, fostering the use of Nature-based Solutions (NbS) within NEB activities and goals. Nevertheless, the transition to the BioCity (Guallart 2020), the BiodiversCities (Zulian et al. 2022) and a decarbonised future requires not only solutions based on functioning ecosystems, as in the case of NbS, but also nature-derived and nature-inspired solutions (IUCN 2020): to the first belong, for example, wind, wave and solar energy and to the second innovative design and material production modelled on biological processes such as biomimicry.

NbS are defined as 'actions to protect, conserve, restore, sustainably use and manage natural or modified terrestrial, freshwater, coastal and marine ecosystems which address social, economic and environmental challenges effectively and adaptively, while simultaneously providing human well-being, ecosystem services, resilience and biodiversity benefits [...]' (UNEA 2022). NbS is an umbrella concept based on previous ecosystem-based approaches grouped into five categories, namely restorative (e.g. ecological restoration), issue-specific (e.g. ecosystem-based adaptation), infrastructure (e.g. natural and green infrastructure), management and protection (Cohen-Shacham et al. 2019). The types of NbS can be also categorised depending on the services provided, the stakeholder groups involved, the engineering levels and the maintenance effort (Eggermont et al. 2015) where ecosystems and human communities (with social, economic and cultural aspects) are at the core.

The uptake of NbS is currently undertaken by developing (1) catalogues and online data pools of existing case studies and best practices (World Bank 2021), (2) tools to support their design and planning (Schröter et al. 2021) and (3) frameworks for their evaluation to measure benefits, co-benefits and costs on the base of indicator values and grades (Raymond et al. 2017) and models forecasting their ecosystem services (e.g. expected improved well-being, reduced temperature and air pollution) under different climatic scenario (Guerry et al. 2023).

In anthropogenic landscapes, such as urban and peri-urban areas, NbS are meant to solve environmental challenges derived by the climatic change and ecosystems' degradation such as heat weaves, floodings, water and air poor quality as well as biodiversity loss, while fostering human well-being, social justice and economic stability (Lafortezza et al. 2018). Along with parks, gardens, street trees and other Green and Blue Infrastructure (GBI) such as bioretention areas and urban farms, green roofs (GRs) and walls act at the urban and neighbourhood scales to tackle the mentioned societal and environmental challenges (Croeser et al. 2021).

The role of biodiversity (genetic, taxonomic, functional, phylogenetic, and habitat) in delivering NbS ecosystem services (ES) is crucial also in the urban context where GRs play an important role, especially in the densely built environment where there is less space for habitats and threes on the ground (Knapp and MacIvor 2023). This work will briefly review several ES provided by GRs focusing on the habitat provisioning of extensive (EGR), simple-intensive (SIGR) and biodiverse green roofs (BGR) as well as the methods and approaches used to implement their biodiversity. The information related to GRs ES is taken from studies on ES provided either by GBI – thus, disentangling the role of GR – or by GR as a stand-alone solution (research focusing either on single or multiple ES provided by GRs).

8.2 Green Roofs: A Nature-Based Solution at the Building Scale

GRs, known also as green rooftops, vegetated roofs or eco-roofs can be synthetically defined as surfaces detached from the ground covered with growing medium, intentionally vegetated and/or spontaneously colonised (SN 564 312 2013). Grs are relatively small ecosystems that can vary from a few square meters, as in the case of a single-family house, to thousands of square meters as in the case of shopping malls (Fig. 8.1).

It is possible to distinguish four types of GRs (Catalano, et al. 2018) according to the substrate thickness and the habitat hosted (Fig. 8.2a–d), which is extensive (dominated by mosses, succulents, annual forbs and grasses), simple-intensive

Fig. 8.1 Examples of green roofs at different scales in Basel (Switzerland): (**a**) extensive green roof of a single-family house of about 100 m^2 (Author's photo). (**b**) Biodiverse green roof of the Stücki Shopping centre of about 3 hectares (Author's photo)

Fig. 8.2 Examples of different types of green roofs in Switzerland: (**a**) the extensive green roof of the ex-Swisscom building in Giubiasco (Author's photo), (**b**) the semi-intensive green roof of the water filtration plant in Wollishofen, Zurich (Author's photo), (**c**) intensive green roof of the Toni Areal campus in Zurich (Author's photo), (**d**) biodiverse green roof of the Baufeld C in Zurich (Author's photo)

(meadow-like habitats with perennial forbs and scattered woody shrubs), intensive (functioning like parks with meadows, shrubs, trees and space for recreation) and biodiverse (meant to replicate specific habitats patches of meadows, brownfields with no vegetation or scattered herbs), which are known as 'biosolar' roofs (Fig. 8.3a, b) when integrated with photovoltaic panels (Catalano and Baumann 2017).

The roots of modern GR technology date to the end of the 1800, when flat roofs covered with a sandy gravel layer protecting the waterproof membrane were eventually colonised by mosses and grass communities ultimately evolving into meadows within the lifespan of the roof (Thommen 1986). These roofs might be

Fig. 8.3 Examples of Biosolar green roofs in Switzerland: (**a**) Hallenbad building in Muttenz (Author's photo); (**b**) Stadtwerk KVA building in Winterthur (Author's photo)

considered the ancestor of EGRs and SIGRs considering that at the beginning of the 1990s, experimental prototypes inspired by the naturally evolving yet spontaneous roofs were vegetated on purpose to test their hydraulic, thermal and horticultural performances (Thuring and Grant 2016). This development led to the publication of the first embryonal German guidelines 'principles for green roofs' in 1982 and eventually to the first guidelines 'for the planning, construction and maintenance of green roofing' in 1990 (FLL 2018; Krupka 1992).

These constructed biotopes at the interface among architecture, landscape architecture and urban ecology, deliver several ES such as regulating (water management, air quality, microclimate), provisioning (biodiversity, pollinator, food when intensive GR is used for urban agriculture) and cultural (community and social benefits) depending on their design and maintenance effort as stand-alone solutions (Berardi et al. 2014; Francis and Jensen 2017; Liu et al. 2021), or in connection with other GBI (Chatzimentor et al. 2020; Marando et al. 2019; Saaroni et al. 2018).

8.2.1 Water Management

In the last decade, GRs were promoted in synergy with rain gardens, bioswales, detention ponds or cisterns thanks to approaches like Sustainable Drainage Systems (SuDS) (Charlesworth et al. 2003), Water Sensitive Design (WSD) (Wong 2006) and, more recently, Sponge City (SC); the latter arose to address solutions combining green and grey infrastructure to store, offset, recycle and purify urban water runoff but also to create a network of multifunctional drainage systems integrated with natural water bodies (Lashford et al. 2019).

GRs behave as a sustainable water management system by reducing the pick of runoff, which otherwise would be conveyed in the canalisation simultaneously by all the impervious surfaces, overloading the sewage systems (Berndtsson 2010). Moreover, GRs allow the water circle to function as it gives back part of it through evapotranspiration (Cascone et al. 2019) while contributing to its quality by reducing the atmospheric pollutants as it passes through the substrate before going to roofs gutters and pipes and then to streams and lakes (Gregoire and Clausen 2011).

Water management services provided by GRs depend on the plant species composition, plant strategy to uptake water (Schrieke et al. 2023), substrate type and thickness (water storage capacity), as well as the intensity of the rain event (Simmons et al. 2008).

8.2.2 *Improved Air Quality*

In 2021, 97% of the European urban population was exposed to levels of PM higher than the guideline level set by the WHO (EEA 2022). The causes of airborne pollution are mainly connected to anthropic activities and vary by pollutant (EEA 2023). For example, the energy production linked to the residential, commercial, and institutional sectors is the main source of particulate matter (PM_{10} and $PM_{2.5}$ and black carbon), and in minor part of nitrogen oxides (NO_x), the latter being mostly due to the road transport sector (Fig. 8.4). Urban GI is considered one solution to improve air quality by exploiting the leaves' ability to trap and remove airborne pollutants.

Regarding the role of GRs in the reduction of air pollution, there are relatively few field experiments on EGRs testing their carbon sequestration potential (Getter et al. 2009; J. Li et al. 2010) and urban particulate absorption (Speak et al. 2012); the majority being the results of simulations using dry deposition models (Currie and Bass 2008; Yang et al. 2008).

The carbon sequestration is performed directly by the plants and substrates (Getter et al. 2009) and indirectly by energy savings due to the reduction of the building energy consumption (Susca 2019) both depending on the substrates' type and depth as well as the plant species and functional diversity (e.g. grasses, bushes, succulents), density and cover. GR performances vary along with species, for example, succulents (e.g. *Sedum* spp.) perform worse than other vascular plants in sequestering carbon (Getter et al. 2009), while grasses (i.e. *Agrostis stolonifera* and *Festuca rubra*) outperform forbs (*Sedum album* and *Plantago lanceolata*) to uptake airborne particulate matter (Speak et al. 2012).

GRs act also as a filter for O_3 (52% of the total), NO_2 (27%) and SO_2 (7%) reaching the annual removal per hectare of 85 kg of pollutants with a higher peak in May when the leaves were fully expanded (Yang et al. 2008). In general, plants

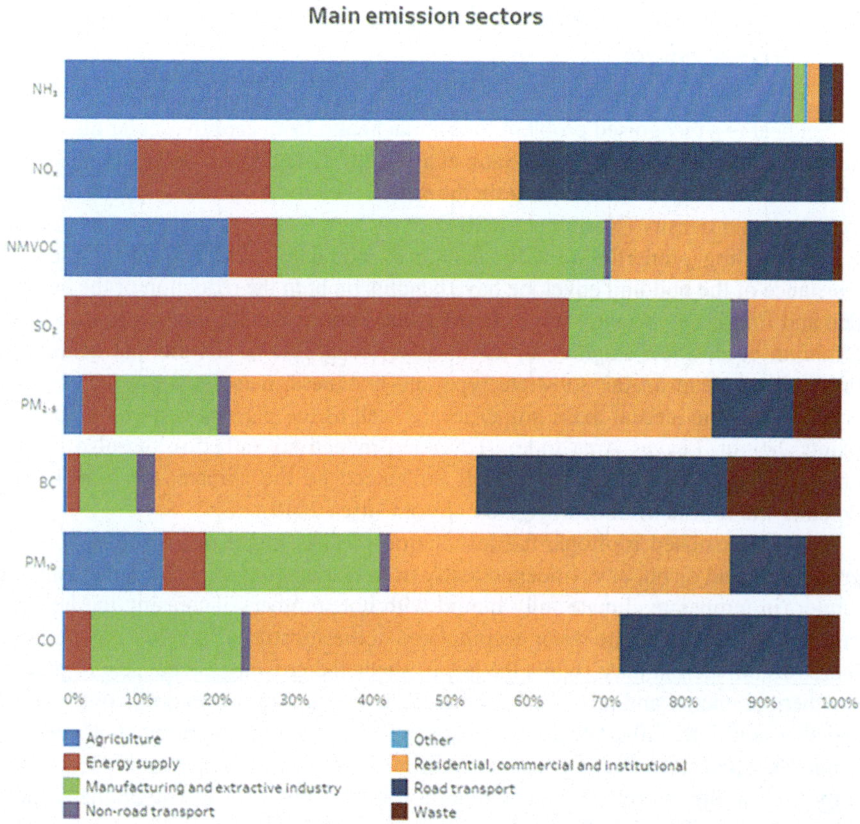

Fig. 8.4 Contributions to EU Member States' emissions of NH3, NO2, NMVOCs, SO2, PM2.5, PM10, BC and CO from the main source sectors in 2021. (Source: from EEA 2023)

(such as perennial herbaceous species) with high photosynthetic capacity, stomatal conductance and transpiration, as well as species with high stomata density, perform better than others in absorbing CO_2 and other gaseous pollutants, while species with hairy, waxy and rough leaves perform better to retain PM (Baraldi et al. 2019). To better contribute to reducing PM, GRs shall be positioned at less than 10 m height and have a plant cover ranging between 50% and 70% (Viecco et al. 2021).

Therefore, if GRs were applied on all the flat roofs of industrial, commercial and contemporary residential areas, they could work as a significant tool for air pollution abatement, especially in the dense urban matrix, where it is not possible to plant street trees but to retrofit/build a high number of flat roofs into GRs (Getter et al. 2009; Li et al. 2010; Rowe 2011).

8.2.3 *Reduction of Building Energy Consumption and Urban Heat Island*

Urban heat is a recognised problem for human health from cardiovascular disease to mental health and mortality (Ode Sang et al. 2022). Urban green spaces mitigate the urban heat island (UHI) depending on the spatial location, vegetation type and urban geometry and density (Norton et al. 2015).

The cooling properties of GRs due to evapotranspiration and added thermal resistance of the building envelope have benefits on both the reduction of the energy demand for air conditioning (both heating and cooling) and the mitigation of UHI (Abhijith et al. 2017; Pugh et al. 2012; Sailor et al. 2012). GR thermal insulation properties depend on the substrate type and thickness, namely, the water storage capacity and the inertial mass properties, as well as the plant selection, because of plants' height, Leaves Area Index (LAI) and reflectivity (albedo), but also leaves emissivity and stomatal resistance, all influenced by the climate, the season and whether the roof is insulated/irrigated or not (Susca 2019).

On non-insulated rooftops, the installation of GRs might decrease the cooling energy demand on hot days (independently from climate) but might be detrimental in winter (in temperate climate fully humid with hot summer), increasing the heating energy demand due to the water accumulated in the substrate. The growing medium increases the inertial mass, while the leaves shade the growing media, absorb part of the thermal energy and provoke heat thermal exchange by convection, contributing together with the substrate to evapotranspiration cooling (Berardi et al. 2014). Plants' positive contribution to the cooling effect depends on their biomass, not only their ability to reflect solar radiation (albedo) but also to evapotranspiration (Blanusa et al. 2013; Lundholm et al. 2010). On self-sustaining roofs (not irrigated), GR performance showed changes throughout seasons due also to the dominant life strategy where annuals die out during summer to sprout the next year (Zheng et al. 2022).

GRs contribute to the reduction of the urban heat island (UHI) regardless of the climate yet show higher potential in dry climates rather than in hot-humid, by reducing air and roof surface temperatures (Jamei et al. 2021). At the pedestrian level, as well as in the case of their effect on air quality, the lower the building and the aspect ratio (building height to street width), the higher the effect that GRs have on reducing UHI and improving outdoor human comfort (Ng et al. 2012; Peng and Jim 2013; Zhang et al. 2019). Moreover, heterogeneous planting with grasses and creeping herbs was shown to perform better than monocultures with *Sedum* spp. due to the higher plant coverage and the species reseeding and resprouting capacity (Robbiati et al. 2022).

In general, the higher the soil moisture, the higher the cooling effect due to the surface and near-surface temperature reduction comparable to high albedo roofs (D. Li et al. 2014). The moisture can be manipulated with irrigation (Kaiser et al. 2019) or as in the case of blue-green roofs technology (known also as multi-layer

Fig. 8.5 An experimental blue-green roof on the building of the engineering department of the University in Palermo (Italy) investigated by Pumo et al. (2023) (Author's photo)

GRs) by adding a layer of water storage underneath the substrate (Fig. 8.5), which showed lower and less fluctuating surface temperatures than gravel (Föllmi et al. 2023) and paved roofs (Pumo et al. 2023).

8.3 Green Roofs as Habitats

Besides providing several environmental benefits, GRs are also small ecosystems (Sutton 2015) constituting habitats for plants (Catalano, Marcenò et al. 2016a, b; Thuring and Dunnett 2014; van Mechelen et al. 2015), microorganisms (Rumble et al. 2018; Schrader and Böning 2006) and other animals such as beetles (Starry et al. 2018), birds (Fernandez-Canero and Gonzalez-Redondo 2010) and bees (Kratschmer et al. 2018).

Ecological studies focused on green roofs' vegetation were conducted mostly in German-speaking countries, where the technique developed into the modern GRs (Thuring and Grant 2016). The first known study dates back to the end of the Second World War, when Kreh (1945) surveyed spontaneous colonised gravel roofs in Stuttgart (Germany) categorised by functional groups, namely, bryophytes, CAM (Crassulacean Acid Metabolism) species and therophytes; substrate depth (ranging between 5 and 20 cm) preferences; pollination and dispersal strategies.

Studies conducted in central Europe described the recurrent plant communities following phytosociological classification (Dengler et al. 2008), thriving on different depths and kinds of substrate (Borchardt 1994; Darius and Drepper 1984; Thommen 1986): stress-tolerant species (of the *Sedo-Scleranthetea* class) can be found on 5–8 cm gravel roofs, while ruderal (*Artemisietea vulgaris* and/or *Stellarietea mediae*) and competitive species (*Molinio-Arrhenatheretea* and *Festuco-Brometea*) at greater depths (Bornkamm 1961; Bossler and Suszka 1988). These studies carried out in central Europe were also confirmed on some flat roofs built at the beginning of 1900 in Trieste (Italy), where the phytocoenosis colonising 15–20 cm of sandy gravel substrate (meant to protect the membrane) belonged mostly to the *Sedo-Scleranthetea* class populated with typical species of extreme habitats such as cliffs (Martini et al. 2004).

Substrate-plant species relationship, including plant-microbial interaction, is crucial for the GRs resilience (plant species survival and dynamic) and performance over time (Chenot et al. 2017; Fulthorpe et al. 2018). Nowadays, the most used substrates are lightweight to avoid extra load on the roofs and the building structure, and it is composed of a mixture of inorganic materials such as lava rock and pumice, expanded shale, clay and slate, as well as organic amendments such as peat and compost, fulfilling standard physicochemical parameters (Eksi et al. 2020). This substrate type can host similar communities as the one found on gravel roofs dominated by *Sedum* spp., winter annuals and few perennials (Thuring and Dunnett 2014). Time (age) is a significant variable influencing soil formation (Schrader and Böning 2006), the kind of pedo-fauna (Rumble et al. 2018) and the plant dynamic on GRs. Old GRs experience shrinking substrate depths, which in turn determine less biomass, lower pH, and greater organic content (Thuring and Dunnett 2014).

GRs are dynamic ecosystems where the humus accumulation, the nutrient supply and the water holding capacity (determined by substrate type and thickness) and the roof height and area are the main drivers for plant community dynamics and functional diversity over time (Catalano Marcenò et al. 2016a, b; Ksiazek-Mikenas and Köhler 2018). These effects are also visible in short-time studies, for example, over 4 years, substrate thickness varying from 5 to 15 cm in single plots showed a greater species richness in the latter (least stressful) but higher functional diversity and evenness in the first (most stressful), whilst a mixed experiment with heterogeneous substrate varying from 5 to 15 cm had toller plants and different plan community than the 10 cm substrate thickness experiment (Heim and Lundholm 2022). Shade is also positively influencing plant diversity on GRs (van der Kolk et al. 2020) as in the case of blue-green roofs where the water storage under the substrate reduces drought stress (van der Kolk et al. 2023).

8.3.1 Biodiverse Green Roofs

The biodiversity of green roofs can be implemented by design in terms of plant species selection, substrate used, and shade (Coulibaly et al. 2023). The so-called 'biodiverse' GRs (BGRs) or 'brown roofs' were developed, respectively, in Switzerland and the United Kingdom to create habitats with biodiversity conservation potential, able to host rare plants and account for animals' life cycles (Baumann 2006; Brenneisen 2006; Grant 2006).

The uptake (both knowledge and construction) of BGRs in Basel and London (Brenneisen and Gedge 2012) was supported by direct incentives to build new GRs or to transform roofs from grey to green, by norms and guidelines and by law requirements. In the city of Basel, for example, the energy department, funded by the Basel Energy Saving Fund, initiated in 1995–1997 the first subside campaign (about one million CHF) sponsoring with 20.- CHF/m^2 for the retrofitting and the construction of new green roofs. In 2005–2007, a second funding campaign (about 1.5 million CHF) sponsored 40.- CHF/m^2 the retrofit of existing roofs (Brenneisen 2010). In 2002, between the two campaigns, a building and construction code made mandatory the greening of every new and retrofitted flat roof following biodiversity-sensitive design guidelines (e.g. use of local substrate, a minimum of 10 cm thickness, selection of local plants and seeds, provision of habitat for invertebrates). The outstanding result of both campaigns and building code was that, in 2010, the city of Basel had almost 100 hectares of green roofs, starting from the 10 hectares before the first campaign (Brenneisen and Gedge 2012).

Keys design features distinguishing BGRs from conventional GRs can be synthesised in (Catalano et al. 2016a): (1) spatial heterogeneity, (2) use of autochthonous plant species and (3) low maintenance and disturbance.

As for the first point, fine-grained patchwork of different, contiguous habitats capable of hosting different biocoenoses can be obtained by varying the substrate thickness and kind (Fig. 8.6a), by adding extra biodiversity-aiding structures (Fig. 8.6b) such as stones, trunks, branches to shelter against weathering the

a b

Fig. 8.6 Biodiversity key design features: (**a**) varying substrate thickness and type on the biodiverse green roof of the Jacob Burckhardt Haus in Basel (Author's photo); (**b**) woodpiles on the biodiverse green roof of the Primarstufe (Kindergarten und Primarschule) in Muttenz (Author's photo)

microfauna and to create different micro-climatic conditions (providing shading), temporary ponds to offer water source for insects and birds and favour the establishment of ephemeral hygrophilous biocenoses linked to temporary wetlands, as well as fostering the survival for ground-nesting birds.

As for the second point, the use of plant species belonging to the regional species pool and therefore already adapted to local conditions enhances the resilience of these novel ecosystems. In this way, green roofs can be part of the greater ecologic network as they can host metapopulations of targeted species that otherwise would not survive in urban environments. Moreover, nurseries would be encouraged to produce seeds and plants of native species.

As for the third point, considering that to moderate disturbance often corresponds to a higher biodiversity, these roofs only need low yearly maintenance (mowing once a year in case of established grasslands, eradication of unwonted seedlings and saplings of woody species which might be eventually too heavy for the roof structure). However, the low maintenance regime required for these kind of GR systems should not influence the periodical check-up of the technical and structural parts.

On BGRs with no irrigation, the most limiting factor for plant growth is the repeated stress related to drought. On such heterogeneous substrate patches, the coarser plants appear to keep water for longer acting eventually as refugia for neighbouring plants growing on more fertile ones (Bates et al. 2013), while intermediate soil depts favour ruderal vegetation and higher depts competitive species (Dunnett 2015). When biodiversity conservation is concerned, the use of local soil (topsoil) is also encouraged (Best et al. 2015) because of the seed bank, microorganisms and fungi contained in it (Fig. 8.7). One good example in this regard is the 100-year-old GRs of the Wollishofen water filtration plant in Zurich where orchid meadows, relict of the past and lost vegetation of the area are still thriving (Landolt 2001). In the short-term, soil-based green roofs showed to have higher biomass and species richness, hosting more spontaneous and native plant species than conventional GRs with mineral-based substrate (van der Kolk et al. 2023).

8.3.2 Plant Species Selection and Design Approaches

Plant species selection is crucial in the design of BGRs. By studying the ecology of spontaneous and old green roofs, Lundholm (2006) introduced the habitat template approach (HTA) aiming at finding habitat analogues (Lundholm and Richardson 2010) to be imitated on GRs. The hypothesis behind this is the similarity in terms of environmental conditions (both climatic and edaphic) between natural habitats and man-made ones (novel habitats). Thus, this approach can be adopted to create near-natural patterns, not only in terms of spatial heterogeneity and substrate properties but also in terms of plant species selection, for example, stress-tolerant species typical to habitats subject to environmental stresses comparable to those imposed by urban ecosystems, like summer drought and periodical floods.

Fig. 8.7 Spontaneous soil-based green roof in Manarola (Italy) (Author's photo)

In southern France, following the concept of the HTA, van Mechelen et al. (2014) suggested a pool of 142 species suitable to grow on green roofs in a Mediterranean climate. The plant list was obtained from vegetation surveys in local grasslands with shallow soils and limestone pavements but also from published phytosociological surveys of the region. The results were refined according to specific functional traits (Raunkiær's life forms and Grime's plant strategies – CSR) obtaining a list of several hemicryptophytes (perennial herbs with buds at soil level), few therophytes (annual plants) and geophytes (perennial herbs with underground buds).

In Italy, Caneva et al. (2015) proposed a list of 138 species including species known to work well on GRs and the following habitat analogues: (1) rocks and screes, (2) grey dunes, (3) perennial grasslands and (4) anthropogenic habitats. The final species list was obtained by applying filters related to chorology, life forms and ecological traits (viz. Ellenberg indicator values) concerning the Italian vascular plants (Guarino et al. 2010; Pignatti et al. 2005). Quite surprisingly, the paper by Caneva et al. (2015) excluded annual and biennial species (therophytes and short-lived hemicryptophytes), which represent a distinctive feature of Mediterranean landscapes, especially grasslands (Guarino et al. 2020), and proved to perform well on green roofs (Vannucchi et al. 2018).

Going beyond species lists, Catalano et al. (2013) proposed a plant sociological approach to select habitats for green roofs. More in detail, they explicitly referred to the two ranks of the sociological hierarchic system, that is, classes and alliances, to create ad hoc seed mixtures based on real plant species assemblages occurring in natural habitats.

Existing GRs can be transformed into BGRs by adding extra features and varying the substrate depths where the roof structure would allow it. In this regard, Catalano et al. (2021) proposed a method to select the most appropriate plant species assemblage, by combining the inductive method proposed by Caneva et al. (2015) and the phytosociological approach. The first step consists of surveying the plants spontaneously growing on the roof to get hints on the environmental conditions by using ecological indicator values (EIVs) (Landolt et al. 2010) related to moisture (F), soil reaction (R), temperature (T), nutrients (N) and light (L). The results are then used to screen the species from existing plant databases by queries related to soil (reaction, moisture and nutrients) and environmental conditions (light and temperature). Finally, the plant species obtained can be grouped according to their occurrence in nature in certain plant communities (phytosociological classes and alliances) to know 'what to sow or plant together?'. Instead, to answer the question 'Where to sow or plant species together?' one option is to run microclimatic simulations or simple shadow analysis to identify the mostly shaded, half-shaded or fully lit surfaces.

In the United Kingdom, Nash et al. (2019) proposed the ecomimicry approach for the design of GRs where for Ecomimicry is meant 'the practice of designing socially responsive and environmental responsible technologies for a particular locale based upon the characteristics of animals, plants, and ecosystems of that locale' (Marshall 2007, p. 2). This approach is based on the 'habitat heterogeneity hypothesis' and ecological restoration practice, the first correlating different niches and resources with higher diversity and the second abiotic conditions, such as climate and microclimate as well as the surrounding habitats, with the species assembly rules and the functioning of the system. For these reasons, the first step is to read the local landscape in terms of soils and substrates, topography, hydrology and aspect, as well as regionally valuable plant and faunal communities.

Independently from the context, the use of low-nutrient substrates (with low organic content) is recommended to favour diverse plant communities, promoting the use of no-commercial standardised substrates as alternative aggregates. Nevertheless, also commercial substrate can boost biodiversity. In this regard, Schröder and Kiehl (2020) tested native seed mixture and racked material from sandy dry grassland, which included lichens and mosses, on commercial substrate for GRs with low content of organic matter. Over 4 years, the sown species performed well in terms of germination and cover especially in the first 2 years. Nevertheless, after an exceptional drought, the cover dropped in all the plots but to a lesser extent in the plots with racked material where cryptogams appeared to have alleviated the severe stress.

8.4 Concluding Remarks

Green roofs are an example of NbS that can provide a wide range of environmental, economic and social benefits, but their effectiveness depends on factors such as climate, building design and the types of plants used. Therefore, the design and implementation of green roofs must be tailored to specific local conditions and goals. Even if the ability to support biodiversity is becoming essential in the implementation of GRs, the relationship between biodiversity and the other co-benefits is still scarce. To make improvements in this sense, designed experiments resulting from a partnership between ecologists and urban designers could develop ecological experiments as socially and politically desirable projects (Felson and Pickett 2005). This kind of collaborative and co-developed nature-based experiments might foster the uptake of BGRs by improving the synergy between research, the industry sector, policymakers and society, towards a mutual exchange rather than trade-offs. While substantial progress has been made in integrating green roofs into urban planning and green buildings, the transition to mainstream adoption requires further commitment from governments, industry stakeholders and the broader community. For this, policies and incentives that encourage the installation of green roofs, coupled with educational initiatives to raise awareness of their benefits, will be essential in supporting their widespread adoption.

Acknowledgements The author acknowledges the support received from the National Recovery and Resilience Plan (NRRP), Mission 4 Component 2 Investment 1.4 - Call for tender No. 3138 of 16 December 2021, rectified by Decree n.3175 of 18 December 2021 of Italian Ministry of University and Research funded by the European Union – NextGenerationEU; Project code CN_00000033, Concession Decree No. 1034 of 17 June 2022 adopted by the Italian Ministry of University and Research, CUP, H43C22000530001 Project title "National Biodiversity Future Center - NBFC".

References

Abhijith KV, Kumar P, Gallagher J, McNabola A, Baldauf R, Pilla F, Broderick B, Di Sabatino S, Pulvirenti B (2017) Air pollution abatement performances of green infrastructure in open road and built-up street canyon environments – a review. Atmos Environ 162:71–86. https://doi.org/10.1016/j.atmosenv.2017.05.014

Baraldi R, Neri L, Costa F, Facini O, Rapparini F, Carriero G (2019) Ecophysiological and micromorphological characterization of green roof vegetation for urban mitigation. Urban For Urban Green 37:24–32. https://doi.org/10.1016/j.ufug.2018.03.002

Bates AJ, Sadler JP, Mackay R (2013) Vegetation development over four years on two green roofs in the UK. Urban For Urban Green 12(1):98–108. https://doi.org/10.1016/j.ufug.2012.12.003

Baumann N (2006) Ground-nesting birds on green roofs in Switzerland: preliminary observations. Urban Habitats 4(1):37–50

Berardi U, GhaffarianHoseini A, GhaffarianHoseini A (2014) State-of-the-art analysis of the environmental benefits of green roofs. Appl Energy 115:411–428. https://doi.org/10.1016/j.apenergy.2013.10.047

Berndtsson JC (2010) Green roof performance towards management of runoff water quantity and quality: a review. Ecol Eng 36(4):351–360. https://doi.org/10.1016/j.ecoleng.2009.12.014

Best BB, Swadek RK, Burgess TL (2015) Soil-based green roofs. In: Sutton RK (ed) Green roof ecosystems. Springer, pp 139–174. https://doi.org/10.1007/978-3-319-14983-7_6

Blanusa T, Vaz Monteiro MM, Fantozzi F, Vysini E, Li Y, Cameron RWF (2013) Alternatives to Sedum on green roofs: can broad leaf perennial plants offer better 'cooling service'? Build Environ 59:99–106. https://doi.org/10.1016/j.buildenv.2012.08.011

Borchardt W (1994) Pflanzenstandort Flachdach: Pflanzenökologische Faktoren der extensiven Begrünung von Flachdächern [Flat roof habitat for plants. Plants ecological parameters of extensive greening of flat roofs]. Das Gartenamt 5:342–346

Bornkamm R (1961) Vegetation und vegetations-entwicklung auf kiesdächern [Vegetation and vegetation development on gravel roofs]. Vegetatio 10(1):1–24

Bossler S, Suszka B (1988) Spontanvegetation auf Dächern [Spontaneous vegetation on roofs] 37: 209–223

Brenneisen S (2006) Space for urban wildlife: designing green roofs as habitats in Switzerland. Urban Habitats 4(1):27–36

Brenneisen S (2010) From pilot to mainstream: green roofs in Basel, Switzerland. Green roofs for Healthy Cities. https://digitalcollection.zhaw.ch/handle/11475/11328

Brenneisen S, Gedge D (2012) Green roof planning in urban areas. In: Meyers RA (ed) Encyclopedia of sustainability science and technology. Springer, pp 4716–4729. https://doi.org/10.1007/978-1-4419-0851-3_208

Caneva G, Kumbaric A, Savo V, Casalini R (2015) Ecological approach in selecting extensive green roof plants: a data-set of Mediterranean plants. Plant Biosyst 149(2):374–383. https://doi.org/10.1080/11263504.2013.819819

Cascone S, Coma J, Gagliano A, Pérez G (2019) The evapotranspiration process in green roofs: a review. Build Environ 147:337–355. https://doi.org/10.1016/j.buildenv.2018.10.024

Catalano C, Baumann N (2017) Biosolar roofs: a symbiosis between biodiverse green roofs and renewable energy. City Green:42–49

Catalano C, Guarino R, Brenneisen S (2013, October 23) A plant sociological approach for extensive green roofs in Mediterranean areas. In: Cities Alive 11th annual green roof and wall conference, San Francisco, 23–26 October. https://doi.org/10.13140/RG.2.2.21245.49126

Catalano C, Brenneisen S, Baumann N, Guarino R (2016a) I tetti verdi di tipo estensivo: Biodiversità ad alta quota. Reticula 12:1–10

Catalano C, Marcenò C, Laudicina VA, Guarino R (2016b) Thirty years unmanaged green roofs: ecological research and design implications. Landsc Urban Plan 149:11–19. https://doi.org/10.1016/j.landurbplan.2016.01.003

Catalano C, Laudicina VA, Badalucco L, Guarino R (2018) Some European green roof norms and guidelines through the lens of biodiversity: do ecoregions and plant traits also matter? Ecol Eng 115:15–26. https://doi.org/10.1016/j.ecoleng.2018.01.006

Catalano C, Pasta S, Guarino R (2021) A plant sociological procedure for the ecological design and enhancement of urban green infrastructure. In: Catalano C, Andreucci MB, Guarino R, Bretzel F, Leone M, Pasta S (eds) Urban services to ecosystems: green infrastructure benefits from the landscape to the urban scale, 1st edn. Springer, pp 31–60. https://doi.org/10.1007/978-3-030-75929-2_3

Charlesworth SM, Harker E, Rickard S (2003) A review of sustainable drainage systems (SuDS): a soft option for hard drainage questions? Geography 88(2):99–107

Chatzimentor A, Apostolopoulou E, Mazaris AD (2020) A review of green infrastructure research in Europe: challenges and opportunities. Landsc Urban Plan 198:103775. https://doi.org/10.1016/j.landurbplan.2020.103775

Chenot J, Gaget E, Moinardeau C, Jaunatre R, Buisson E, Dutoit T (2017) Substrate composition and depth affect soil moisture behavior and plant-soil relationship on Mediterranean extensive green roofs. Water 9(11):11. https://doi.org/10.3390/w9110817

Cohen-Shacham E, Andrade A, Dalton J, Dudley N, Jones M, Kumar C, Maginnis S, Maynard S, Nelson CR, Renaud FG, Welling R, Walters G (2019) Core principles for successfully implementing and upscaling nature-based solutions. Environ Sci Pol 98:20–29. https://doi.org/10.1016/j.envsci.2019.04.014

Coulibaly SFM, Aubry C, Provent F, Rousset-Rouvière S, Joimel S (2023) The role of green roofs as urban habitats for biodiversity modulated by their design: a review. Environ Res Lett 18(7): 073003. https://doi.org/10.1088/1748-9326/acd801

Croeser T, Garrard G, Sharma R, Ossola A, Bekessy S (2021) Choosing the right nature-based solutions to meet diverse urban challenges. Urban For Urban Green 65:127337. https://doi.org/10.1016/j.ufug.2021.127337

Currie BA, Bass B (2008) Estimates of air pollution mitigation with green plants and green roofs using the UFORE model. Urban Ecosyst 11(4):409–422. https://doi.org/10.1007/s11252-008-0054-y

Darius F, Drepper J (1984) Rasendächer in West-Berlin [Roofs meadow in West Berlin]. Das Gartenamt 33:309–315

Dengler J, Chytrý M, Ewald J (2008) Phytosociology. In: Jørgensen SE, Fath BD (eds) Encyclopedia of ecology. Academic Press, pp 2767–2779. https://doi.org/10.1016/B978-008045405-4.00533-4

Dunnett N (2015) Ruderal green roofs. In: Sutton RK (ed) Green roof ecosystems. Springer, pp 233–255. https://doi.org/10.1007/978-3-319-14983-7_10

EEA (2022) Air quality in Europe 2022. Web report. European Environment Agency. https://www.eea.europa.eu/publications/air-quality-in-europe-2022

EEA (2023) Air pollution in Europe: 2023 reporting status under the National Emission reduction commitments directive. Briefing. European Environment Agency. https://www.eea.europa.eu/publications/national-emission-reduction-commitments-directive-2023

Eggermont H, Balian E, Azevedo JMN, Beumer V, Brodin T, Claudet J, Fady B, Grube M, Keune H, Lamarque P, Reuter K, Smith M, van Ham C, Weisser WW, Le Roux X (2015) Nature-based solutions: new influence for environmental management and research in Europe. GAIA Ecol Perspect Sci Soc 24(4):243–248. https://doi.org/10.14512/gaia.24.4.9

Eksi M, Sevgi O, Akburak S, Yurtseven H, Esin İ (2020) Assessment of recycled or locally available materials as green roof substrates. Ecol Eng 156:105966. https://doi.org/10.1016/j.ecoleng.2020.105966

Felson AJ, Pickett ST (2005) Designed experiments: new approaches to studying urban ecosystems. Front Ecol Environ 3(10):549–556. https://doi.org/10.1890/1540-9295(2005)003[0549:DENATS]2.0.CO;2

Fernandez-Canero R, Gonzalez-Redondo P (2010) Green roofs as habitat for birds: a review. J Anim Vet Adv 9:15

Fetting C (2020) The new European Green Deal (ESDN report). ESDN Office, Vienna

FLL (2018) Dachbegrünungsrichtlinien Richtlinien für Planung, Bau und Instandhaltung von Dachbegrünungen. Forschungsgesellschaft Landschaftsentwicklung Landschaftsbau e. V. https://shop.fll.de/de/downloadable/download/sample/sample_id/133/

Föllmi D, Corpel L, Solcerova A, Kluck J (2023) Influence of blue-green roofs on surface and indoor temperatures over a building scale. Nature-Based Solutions 4:100076. https://doi.org/10.1016/j.nbsj.2023.100076

Francis LFM, Jensen MB (2017) Benefits of green roofs: a systematic review of the evidence for three ecosystem services. Urban For Urban Green 28:167–176. https://doi.org/10.1016/j.ufug.2017.10.015

Fulthorpe R, MacIvor JS, Jia P, Yasui S-LE (2018) The green roof microbiome: improving plant survival for ecosystem service delivery. Front Ecol Evol 6. https://www.frontiersin.org/articles/10.3389/fevo.2018.00005

Getter KL, Rowe DB, Robertson GP, Cregg BM, Andresen JA (2009) Carbon sequestration potential of extensive green roofs. Environ Sci Technol 43(19):7564–7570. https://doi.org/10.1021/es901539x

Grant G (2006) Extensive green roofs in London. Urban Habitats 4(1):51–65
Gregoire BG, Clausen JC (2011) Effect of a modular extensive green roof on stormwater runoff and water quality. Ecol Eng 37(6):963–969. https://doi.org/10.1016/j.ecoleng.2011.02.004
Guallart V (2020) From digital cities to biocities: harnessing the power of the digital revolution to reinvent the urban ecology model. Archit Des 90(3):72–75. https://doi.org/10.1002/ad.2571
Guarino R, Addamiano S, Rosa ML, Pignatti S (2010) "Flora Italiana Digitale": an interactive identification tool for the Flora of Italy. In: Nimis PL, Vignes Lebbe R (eds) Tools for identifying biodiversity: Progress and problems
Guarino R, Vrahnakis M, Rojo MPR, Giuga L, Pasta S (2020) Grasslands and Shrublands of the Mediterranean region. In: Goldstein MI, DellaSala DA (eds) Encyclopedia of the world's biomes. Elsevier, pp 638–655. https://doi.org/10.1016/B978-0-12-409548-9.12119-0
Guerry AD, Lonsdorf EV, Nootenboom C, Remme RP, Griffin R, Waters H, Polasky S, Han B, Wu T, Janke BD, Meacham M, Hamel P, Wang X (2023) Mapping, measuring, and valuing the benefits of nature-based solutions in cities. In: Nature-based solutions for cities. Edward Elgar Publishing, pp 259–293. https://www.elgaronline.com/edcollchap-oa/book/9781800376762/book-part-9781800376762-23.xml
Heim A, Lundholm J (2022) Changes in plant community composition and functional plant traits over a four-year period on an extensive green roof. J Environ Manag 304:114154. https://doi.org/10.1016/j.jenvman.2021.114154
IUCN (2020) Guidance for using the IUCN global standard for nature-based solutions: a user-friendly framework for the verification, design and scaling up of nature-based solutions, 1st edn. IUCN. https://doi.org/10.2305/IUCN.CH.2020.09.en
Jamei E, Chau HW, Seyedmahmoudian M, Stojcevski A (2021) Review on the cooling potential of green roofs in different climates. Sci Total Environ 791:148407. https://doi.org/10.1016/j.scitotenv.2021.148407
Kaiser D, Köhler M, Schmidt M, Wolff F (2019) Increasing evapotranspiration on extensive green roofs by changing substrate depths, construction, and additional irrigation. Buildings 9(7):7. https://doi.org/10.3390/buildings9070173
Knapp S, MacIvor JS (2023) Nature-based solutions and biodiversity: synergies, trade-offs, and ways forward. In: Nature-based solutions for cities. Edward Elgar, pp 82–103. https://www.elgaronline.com/edcollchap-oa/book/9781800376762/book-part-9781800376762-14.xml
Kratschmer S, Kriechbaum M, Pachinger B (2018) Buzzing on top: linking wild bee diversity, abundance and traits with green roof qualities. Urban Ecosyst 21(3):429–446. https://doi.org/10.1007/s11252-017-0726-6
Kreh W (1945) Die Pflanzenwelt unserer Kiesdächer [The plant world of our gravel roofs]. Jahreshefte Des Vereins Für Vaterländische Naturkunde in Württemberg 97(101):199–207
Krupka B (1992) Dachbegrünung. Pflanzen- und Vegetationsanwendung an Bauwerken (1992nd ed.). Verlag Eugen Ulmer
Ksiazek-Mikenas K, Köhler M (2018) Traits for stress-tolerance are associated with long-term plant survival on green roofs. J Urban Ecol 4(1):juy016. https://doi.org/10.1093/jue/juy016
Lafortezza R, Chen J, van den Bosch CK, Randrup TB (2018) Nature-based solutions for resilient landscapes and cities. Environ Res 165:431–441. https://doi.org/10.1016/j.envres.2017.11.038
Landolt E (2001) Orchideen-Wiesen in Wollishofen (Zürich) – Ein erstaunliches Relikt aus der Anfang des 20. Jahrhunderts. Vierteljahrsschrift der Naturforschenden Gesellschaft in Zürich 146(2–3):41–51
Landolt E, Bäumler B, Erhardt A, Hegg O, Klötzli F, Lämmler W, Nobis M, Rudmann-Maurer K, Schweingruber FH, Theurillat J-P, Urmi E, Vust M, Wohlgemuth T (2010) Flora indicativa Ökologische Zeigerwerte und biologische Kennzeichen zur Flora der Schweiz und der Alpen (1., Aufl. edizione). Haupt
Lashford C, Rubinato M, Cai Y, Hou J, Abolfathi S, Coupe S, Charlesworth S, Tait S (2019) SuDS & sponge cities: a comparative analysis of the implementation of pluvial flood management in the UK and China. Sustainability 11(1):213. https://doi.org/10.3390/su11010213

Li J, Wai OWH, Li YS, Zhan J, Ho YA, Li J, Lam E (2010) Effect of green roof on ambient CO2 concentration. Build Environ 45(12):2644–2651. https://doi.org/10.1016/j.buildenv.2010.05.025

Li D, Bou-Zeid E, Oppenheimer M (2014) The effectiveness of cool and green roofs as urban heat Island mitigation strategies. Environ Res Lett 9(5):055002. https://doi.org/10.1088/1748-9326/9/5/055002

Liu H, Kong F, Yin H, Middel A, Zheng X, Huang J, Xu H, Wang D, Wen Z (2021) Impacts of green roofs on water, temperature, and air quality: a bibliometric review. Build Environ 196:107794. https://doi.org/10.1016/j.buildenv.2021.107794

Lundholm J (2006) Green roofs and facades: a habitat template approach. Urban Habitats 4(1):87–101

Lundholm J, Richardson PJ (2010) Habitat analogues for reconciliation ecology in urban and industrial environments. J Appl Ecol 47(5):966–975

Lundholm J, MacIvor JS, MacDougall Z, Ranalli M (2010) Plant species and functional group combinations affect green roof ecosystem functions. PLoS One 5(3):e9677. https://doi.org/10.1371/journal.pone.0009677

Marando F, Salvatori E, Sebastiani A, Fusaro L, Manes F (2019) Regulating ecosystem services and green infrastructure: assessment of urban heat Island effect mitigation in the municipality of Rome, Italy. Ecol Model 392:92–102. https://doi.org/10.1016/j.ecolmodel.2018.11.011

Marshall A (2007) The theory and practice of ecomimicry. Working paper series, 3

Martini F, Codogno M, Comelli L, Mesiano O (2004) La vegetazione dei tetti verdi a Trieste. Fitosociologia 41(1):181–192

Nash C, Ciupala A, Gedge D, Lindsay R, Connop S (2019) An ecomimicry design approach for extensive green roofs. J Living Archit 6(1):1

Ng E, Chen L, Wang Y, Yuan C (2012) A study on the cooling effects of greening in a high-density city: an experience from Hong Kong. Build Environ 47:256–271. https://doi.org/10.1016/j.buildenv.2011.07.014

Norton BA, Coutts AM, Livesley SJ, Harris RJ, Hunter AM, Williams NSG (2015) Planning for cooler cities: a framework to prioritise green infrastructure to mitigate high temperatures in urban landscapes. Landsc Urban Plan 134:127–138. https://doi.org/10.1016/j.landurbplan.2014.10.018

Ode Sang Å, Thorpert P, Fransson A-M (2022) Planning, designing, and managing green roofs and green walls for public health – an ecosystem services approach. Front Ecol Evol 10. https://www.frontiersin.org/articles/10.3389/fevo.2022.804500

Peng LLH, Jim CY (2013) Green-roof effects on neighborhood microclimate and human thermal sensation. Energies 6(2):2. https://doi.org/10.3390/en6020598

Pignatti S, Menegoni P, Pietrosanti S (2005) Indicazione attraverso le piante vascolari. Valori di indicazione secondo Ellenberg (Zeigerwerte) per le specie della Flora d'Italia. Braun-Blanquetia 3:91–97

Pugh TAM, MacKenzie AR, Whyatt JD, Hewitt CN (2012) Effectiveness of green infrastructure for improvement of air quality in urban street canyons. Environ Sci Technol 46(14):7692–7699. https://doi.org/10.1021/es300826w

Pumo D, Alongi F, Cannarozzo M, Noto LV (2023) Climate adaptive urban measures in Mediterranean areas: thermal effectiveness of an advanced multilayer green roof installed in Palermo (Italy). Build Environ 243:110731. https://doi.org/10.1016/j.buildenv.2023.110731

Raymond CM, Frantzeskaki N, Kabisch N, Berry P, Breil M, Nita MR, Geneletti D, Calfapietra C (2017) A framework for assessing and implementing the co-benefits of nature-based solutions in urban areas. Environ Sci Pol 77:15–24. https://doi.org/10.1016/j.envsci.2017.07.008

Robbiati FO, Cáceres N, Hick EC, Suarez M, Soto S, Barea G, Matoff E, Galetto L, Imhof L (2022) Vegetative and thermal performance of an extensive vegetated roof located in the urban heat Island of a semiarid region. Build Environ 212:108791. https://doi.org/10.1016/j.buildenv.2022.108791

Rowe DB (2011) Green roofs as a means of pollution abatement. Environ Pollut 159(8):2100–2110. https://doi.org/10.1016/j.envpol.2010.10.029
Rumble H, Finch P, Gange AC (2018) Green roof soil organisms: anthropogenic assemblages or natural communities? Appl Soil Ecol 126:11–20. https://doi.org/10.1016/j.apsoil.2018.01.010
Saaroni H, Amorim JH, Hiemstra JA, Pearlmutter D (2018) Urban green infrastructure as a tool for urban heat mitigation: survey of research methodologies and findings across different climatic regions. Urban Clim 24:94–110. https://doi.org/10.1016/j.uclim.2018.02.001
Sailor DJ, Elley TB, Gibson M (2012) Exploring the building energy impacts of green roof design decisions – a modeling study of buildings in four distinct climates. J Build Phys 35(4):372–391. https://doi.org/10.1177/1744259111420076
Schrader S, Böning M (2006) Soil formation on green roofs and its contribution to urban biodiversity with emphasis on Collembolans. Pedobiologia 50(4):347–356. https://doi.org/10.1016/j.pedobi.2006.06.003
Schrieke D, Szota C, Williams NSG, Farrell C (2023) Evaluating the effectiveness of spontaneous vegetation for stormwater mitigation on green roofs. Sci Total Environ 898:165643. https://doi.org/10.1016/j.scitotenv.2023.165643
Schröder R, Kiehl K (2020) Extensive roof greening with native sandy dry grassland species: effects of different greening methods on vegetation development over four years. Ecol Eng 145: 105728. https://doi.org/10.1016/j.ecoleng.2020.105728
Schröter B, Zingraff-Hamed A, Ott E, Huang J, Hüesker F, Nicolas C, Schröder NJS (2021) The knowledge transfer potential of online data pools on nature-based solutions. Sci Total Environ 762:143074. https://doi.org/10.1016/j.scitotenv.2020.143074
Simmons MT, Gardiner B, Windhager S, Tinsley J (2008) Green roofs are not created equal: the hydrologic and thermal performance of six different extensive green roofs and reflective and non-reflective roofs in a sub-tropical climate. Urban Ecosyst 11(4):339–348. https://doi.org/10.1007/s11252-008-0069-4
SN 564 312 (2013) Begrünung von Dächern (312). http://shop.sia.ch/normenwerk/architekt/sia%20312/d/D/Product
Speak AF, Rothwell JJ, Lindley SJ, Smith CL (2012) Urban particulate pollution reduction by four species of green roof vegetation in a UK city. Atmos Environ 61:283–293. https://doi.org/10.1016/j.atmosenv.2012.07.043
Starry O, Gonsalves S, Ksiazek-Mikenas K, MacIvor JS, Gardner M, Szallies A, Brenneisen S (2018) A global comparison of beetle community composition on green roofs and the potential for homogenization, Vol. 1, p 15
Susca T (2019) Green roofs to reduce building energy use? A review on key structural factors of green roofs and their effects on urban climate. Build Environ 162:106273. https://doi.org/10.1016/j.buildenv.2019.106273
Sutton RK (ed) (2015) Green roof ecosystems, vol 223. Springer. https://doi.org/10.1007/978-3-319-14983-7
Thommen MS (1986) Pflanzengemeinschaften natürlich besiedelter Kiesdächer und extensiver Dachbegrünungen [Plant communities of naturally colonised gravel roofs and extensive green roofs]. Master thesis. University of Basel
Thuring C, Dunnett N (2014) Vegetation composition of old extensive green roofs (from 1980s Germany). Ecol Process 3(1):4. https://doi.org/10.1186/2192-1709-3-4
Thuring C, Grant G (2016) The biodiversity of temperate extensive green roofs – a review of research and practice. Isr J Ecol Evol 62(1–2):44–57. https://doi.org/10.1080/15659801.2015.1091190
UNEA (2022) Resolution adopted by the United Nations Environment Assembly on 2 March 2022 (UNEP/EA.5/Res.5). United Nations. https://www.unep.org/resources/resolutions-treaties-and-decisions/UN-Environment-Assembly-5-2
van der Kolk H-J, van den Berg P, Korthals G, Bezemer TM (2020) Shading enhances plant species richness and diversity on an extensive green roof. Urban Ecosyst 23(5):935–943. https://doi.org/10.1007/s11252-020-00980-w

van der Kolk H-J, van den Berg P, Veen T, Bezemer TM (2023) Substrate composition impacts long-term vegetation development on blue-green roofs: insights from an experimental roof and greenhouse study. Ecol Eng 186:106847. https://doi.org/10.1016/j.ecoleng.2022.106847

van Mechelen C, Dutoit T, Hermy M (2014) Mediterranean open habitat vegetation offers great potential for extensive green roof design. Landsc Urban Plan 121:81–91. https://doi.org/10.1016/j.landurbplan.2013.09.010

van Mechelen C, Dutoit T, Hermy M (2015) Vegetation development on different extensive green roof types in a Mediterranean and temperate maritime climate. Ecol Eng 82:571–582. https://doi.org/10.1016/j.ecoleng.2015.05.011

Vannucchi F, Pini R, Scatena M, Benelli G, Canale A, Bretzel F (2018) Deinking sludge in the substrate reduces the fertility and enhances the plant species richness of extensive green roofs. Ecol Eng 116:87–96. https://doi.org/10.1016/j.ecoleng.2018.02.027

Viecco M, Jorquera H, Sharma A, Bustamante W, Fernando HJS, Vera S (2021) Green roofs and green walls layouts for improved urban air quality by mitigating particulate matter. Build Environ 204:108120. https://doi.org/10.1016/j.buildenv.2021.108120

Wong THF (2006) Water sensitive urban design – the journey thus far. Aust J Water Resour 10(3): 213–222. https://doi.org/10.1080/13241583.2006.11465296

World Bank (2021) A catalogue of nature-based solutions for urban resilience. World Bank. https://openknowledge.worldbank.org/handle/10986/36507

Yang J, Yu Q, Gong P (2008) Quantifying air pollution removal by green roofs in Chicago. Atmos Environ 42(31):7266–7273. https://doi.org/10.1016/j.atmosenv.2008.07.003

Zhang G, He B-J, Zhu Z, Dewancker BJ (2019) Impact of morphological characteristics of green roofs on pedestrian cooling in subtropical climates. Int J Environ Res Public Health 16(2):2. https://doi.org/10.3390/ijerph16020179

Zheng X, Yang Z, Yang J, Tang M, Feng C (2022) An experimental study on the thermal and energy performance of self-sustaining green roofs under severe drought conditions in summer. Energ Buildings 261:111953. https://doi.org/10.1016/j.enbuild.2022.111953

Zulian G, Marando F, Vogt P, Vignola GB, Almenar JB, Zurbarán-Nucci M, Prince K (2022) BiodiverCities: a roadmap to enhance the biodiversity and green infrastructure of European cities by 2030. Publications Office of the European Union. https://data.europa.eu/doi/10.2760/21172

Chapter 9
The New Special Collserola Range Protection Plan [PEPNat]: A Natural Park Within the Metropolitan Area of Barcelona

Eugènia Vidal-Casanovas

Abstract The Collserola mountain range is a singular natural area covering over 8000 ha at the heart of a complex metropolitan reality. It thus has great potential in terms of providing ecosystem services on a metropolitan and regional level, yet little ecological continuity with other nearby open spaces. As a result of the creation of Collserola Natural Park in 2010, a new special protection plan, the PEPNat, has been formally approved (April 2021). This Plan aims to preserve biodiversity and increase ecosystem services under dynamic and adaptive management. The presentation focuses on the holistic and comprehensive approach of the new plan in the frame of the metropolitan green infrastructure.

Keywords Barcelona · Collserola Natural Park · Comprehensive approach · Systemic vision

9.1 The Metropolitan Area of Barcelona as an Interconnected Green Metropolis

9.1.1 Open Spaces Within the Metropolis

The metropolis is a new kind of urban form that goes beyond size and scale. This concept emerged in the early twentieth century, when European and American cities expanded beyond their previous scale, spreading out in complex forms of

An article based on the PEPNat and MPGMCo documentation definitively approved by the Government of Catalonia in 2021 and 2020, respectively. The documents were prepared by the AMB Technical Service. They can be accessed at the following link: https://www.amb.cat/es/web/territori/urbanisme/pepnat-i-mpgmco. The Plan received the Manuel de Solà-Morales Urban Planning Award in 2022.

E. Vidal-Casanovas (✉)
Green Infrastructure Service, Barcelona Metropolitan Area [OTG-AMB], Catalonia, Spain
e-mail: evidal@amb.cat

F. Dal Falco (ed.), *Beyond the Garden*, Designing Environments,
https://doi.org/10.1007/978-3-031-50567-6_9

occupation, with activities separating in an ever-changing mix of building types. This included diverse urban centres, peripheries and city-dependent spaces that are extensively and spatially connected (Solà-Morales 2008).

Open spaces occupy different positions and perform multiple roles in this new urban form. In addition to the peripheral position, linked to the modern concentric territorial model, natural and semi-natural protected areas are now located beyond, between and within different types of occupation of urban land. This might make them not only more fragile but also more decisive.

As cities have grown and expanded, protected areas have often become fragmented and isolated, with a loss of habitat and a decline in ecological connectivity due to development, infrastructure and other urban pressures. Moreover, metropolitan areas are major sources of pollution and other negative impacts. Furthermore, climate change is expected to have significant impacts on these areas, including altered ecosystems, changes in species distribution and an increased risk of wildfires.

Meanwhile, the proximity and constant interrelationship between urban fabrics and open spaces increases the potential of these spaces as providers of multiple environmental functions – such as the scupper effect of vegetation, conservation of the water cycle and air quality and economic and social functions (both tangible and intangible).

9.1.2 The Metropolitan Area of Barcelona

The metropolitan area of Barcelona as a new urban form has developed over the last century, resulting in a territorial, social, demographic, economic and cultural entity. Today, it encompasses 36 municipalities and 3.2 million inhabitants over 636 square kilometres, comprising various urban settlements and a dense network of infrastructure and services. As the largest metropolitan conurbation in the western Mediterranean, it accounts for half of Catalonia's GDP.

However, despite its density and complex metropolitan dynamics, open spaces occupy 54.6% of the metropolitan area of Barcelona, and the vast majority of them are protected (more than 90%) (AMB SRPD 2019b: 7). The topography has made it possible to preserve spaces with high ecosystem service value, such as the Collserola range, the Garraf massif, the Ordal mountains and the Marina mountain range (Fig. 9.1). Most of these spaces have retained their forest cover, which combines with farming activities to form an agroforestry mosaic. The metropolitan area also has other important natural and semi-natural spaces, such as the Llobregat and Besòs rivers, the natural areas surrounding the Llobregat delta, the Agrarian Park and more than 30 km of beaches and a network of over 50 urban and peri-urban parks. All these spaces make up a highly diverse green infrastructure, which plays a key role in the future of the metropolis and which is full of environmental, productive and social assets. This green infrastructure accounts for more than 70% of the territory (AMB SRPD 2019b: 7).

Fig. 9.1 Serra de Collserola Natural Park and the metropolitan green infrastructure. (Source: AMB)

9.2 The New Collserola Range Protection Plan (PEPNat)

Located at the heart of this complex metropolitan context, the Collserola mountain range is a unique natural area covering over 8000 ha, taking in nine different municipalities (Fig. 9.2). Situated in the middle of a densely populated and congested area, the Park is today a well-preserved biodiverse territory of farmland and forests, with the potential to play a key role in the environmental and ecosystem balance of the metropolitan area (Vidal-Casanovas et al. 2022).

The creation of the Serra de Collserola Natural Park in 2010 gave rise to the current special protection plan, PEPNat, which aims to preserve biodiversity and increase ecosystem services under dynamic and responsive management. The new special plan is a hybrid plan, which combines environmental and urban planning aspects and replaces the PEPCo, the plan that had been in force since 1987. PEPNat was drafted by the Barcelona Metropolitan Area (AMB) within a framework of co-governance and very broad consensus, under the guidance of a technical and institutional commission formed by the Serra de Collserola Natural Park Consortium and the administrations involved (the Government of Catalonia, AMB, Barcelona Provincial Council and the nine municipalities contained in the Park) and chaired by the Director General for Environmental Policies and Natural Environment of the Government of Catalonia.

It covers an area of 8156 ha that also includes the partial nature reserve of La Rierada–Can Balasc (383 ha) and La Font Groga (117 ha). In addition to the land declared part of the Natural Park, PEPNat contains a functional space consisting of areas that, due to their characteristics or functionality, are fundamental in attaining the objectives of protecting and improving the biodiversity of the Park. The total area of the functional space, which includes both interior and perimeter areas, is 3676 ha.

Fig. 9.2 View of the metropolitan area from Passeig de les Aigües in the Collserola Park. (Source: AMB)

The most important innovation in the Plan is the paradigm shift it embodies. PEPNat introduces a new vision for the protection of natural spaces that is particularly groundbreaking from an urban planning perspective. Unlike the previous model, which was based on a snapshot of specific landscape types and units and which was a response to the characteristics of the vegetation and the territory at the time the Plan was drafted, PEPNat recognises the processes and functions that take place within the Park and beyond. Apart from this fundamental change in understanding the protection of natural spaces, other significant aspects of the Plan are the comprehensive approach and the systemic vision inherent in the Plan, which are two interrelated key aspects in the planning of metropolitan open spaces.

9.2.1 Comprehensive Approach

Environmental challenges, especially in complex urban contexts such as the metropolitan area of Barcelona, cannot be addressed by a single discipline, but instead require collaboration and integration of various fields of knowledge for a more complete and effective understanding of these socio-ecological systems. As stated in the credits of the Plan, many technicians from different disciplines and government bodies have been involved in its drafting. Table 9.1 summarises the topics that were taken into account when this multidisciplinary team considered the state of the Park.

Table 9.1 Diagnosis and general objectives of PEPNat

Diagnosis	General objectives
Ecology and preservation of biodiversity	Maintain and improve the environmental conditions of the Park and conserve its biodiversity, habitats and ecological processes
Edge spaces	Focus on the outer limits of the Park to reduce risks, ensure ecological and social connectivity, and improve its urban character
Social use model	Sustainably manage the social and recreational use of the Park within a metropolitan context
Valorisation of natural resources	Maintain and adapt the activities related to the use of natural resources and promote the green economy
Ecosystem services	Identify and value ecosystem services, furthering research into Collserola as a space for knowledge
Heritage	Maintain and improve the Park's architectural, historical and cultural heritage
Landscape	Modulate the landscape and minimise impacts
Management model	Establish an active and direct management regime based on public-private collaboration models
Regulatory instrument	Support the adoption of an open, flexible regulatory instrument, adaptable to the changing territory and society

Source: Author's elaboration

PEPNat encompasses the management and regulation of their uses, their intensities, their guidelines, their recommendations and their strategic projects that will be implemented by the managing body. In accordance with the new Park model, the wide variety of contributions has made it possible to define and develop an ecological strategy that covers various thematic areas, ranging from the preservation and improvement of ecological values, to the enhancement of natural resources, public use, built heritage, infrastructures and services and the functional space. The sections below summarise the main proposals and new features of the Plan in relation to each of these thematic areas.

9.2.1.1 Preservation and Enhancement of Ecological Values

The proposals presented in this section focus on four main aspects, namely, improving both internal and external ecological connectivity, preserving elements of ecological value, controlling disturbances and promoting the green economy. All the proposals included in the Plan are based around these four areas.

The Plan establishes an environmental zoning system within the general framework of a single zoning code that provides maximum protection, code 29co. In addition to the partial nature reserves that were established by Decree 146/2010, the PEPNat defines four other areas crucial to biodiversity conservation and the enhancement of ecosystem services, namely, patches of connectivity interest, areas of special importance, islands of tranquillity and priority areas for agriculture. These areas are based on specific processes or functions of the Park and play an essential role in its current and future management (Fig. 9.3).

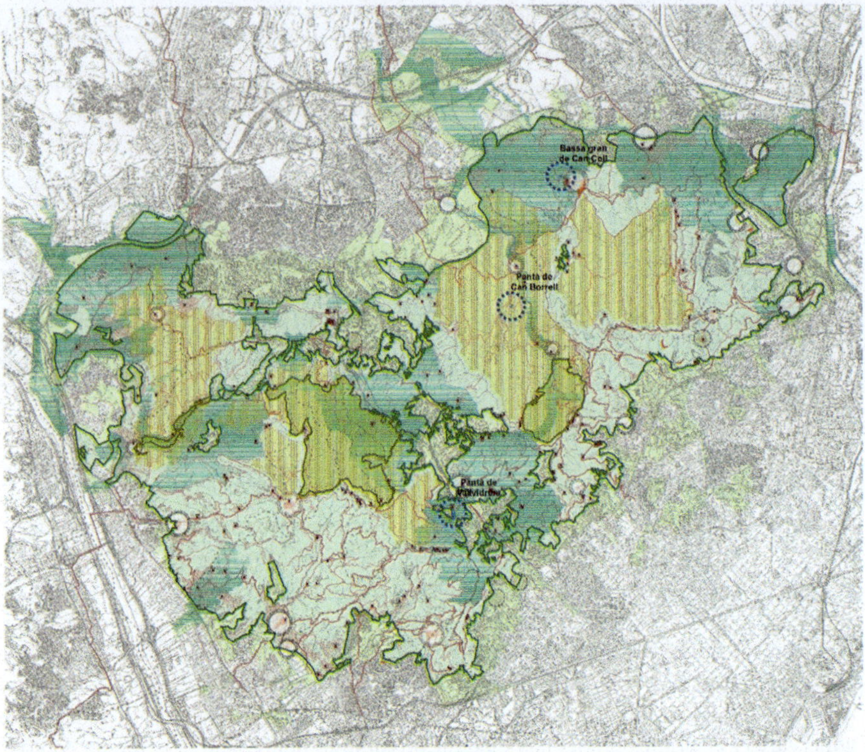

Fig. 9.3 The new Park structure proposed by PEPNat is based on a unique urban planning zoning code where different environmental zoning areas can overlap. The circles and crosses relate to the strategies assigned to the existing recreation areas and traditional old buildings, respectively. (Source: PEPNat)

9.2.1.2 Valorisation of Natural Resources

Forestry, agriculture and livestock farming are among the Park's primary activities. All these activities are permitted within PEPNat throughout the ark, albeit with strong restrictions in the most sensitive areas. One of the Plan's main goals is to increase and restore agricultural land uses, given that the amount of cropland has fallen considerably in recent decades. Nevertheless, recent years have seen an increase in the local demand for quality products and the number of people interested in becoming farmers (Associació Arran de Terra 2017). This also reinforces the goals of promoting the agroforestry mosaic and the correct management of habitats and species.

Given this context, PEPNat encourages the establishment of synergies between the management body and the owners of agroforestry estates. More specifically, it facilitates sustainable agricultural activities, and it establishes a link between the permitted uses of existing traditional buildings and sustainable estate management. This management also fosters the agroforestry mosaic and safeguards the role played

Fig. 9.4 Can Calopa de Dalt is an example of the link between old cultural heritage and active estate management. This *masia* hosts a wine production cooperative and a Special Work Centre (CET) for people with disabilities. (Source: CPNSC)

by patches as stepping stones in the improvement of habitat area, quality and connectivity, among other aspects (Fig. 9.4).

9.2.1.3 Public Use

The Plan establishes a basic network of paths, recreational and educational areas, prioritising the reduction of social pressure in sensitive areas within the Park and promoting non-motorised means of access to the park (Fig. 9.5).

This is consistent with the objectives of PEPNat and takes into account the pre-existing infrastructure for public use, as well as the relationship between the Park and other metropolitan open spaces. The Plan takes advantage of existing infrastructure for public use (paths, recreational areas and public facilities) on one hand while creating islands of tranquillity on the other. The two strategies share the same goal.

The islands are produced by identifying the least disturbed areas, mainly as regards social use of the Park, and any other feature or activity that creates disturbances in ecosystems and their functionality. They are spaces that require favourable conditions for flora and fauna and receive special treatment. Access to these islands is therefore expected to be more restricted, and possible disturbances are subject to specific conditions.

Fig. 9.5 Users of the park along the Passeig de les Aigües. (Source: CPNSC)

9.2.1.4 Built Heritage

The proposals concerning the built heritage of Collserola Park aim to achieve two main objectives: first, to restore and enhance the remaining cultural heritage of the area and, second, to restrict new construction except in some exceptional cases related to primary activities and management of the social use of the Park.

In order to accomplish these goals, a catalogue has been created that lists a total of 199 structures, including *masies* (traditional rural gable-roofed constructions common in eastern Spain), rural homes and other buildings that were built sustainably and in harmony with the local agriculture and natural landscape.

The purpose of this catalogue is to regulate the use and maintenance of the Park's architectural, historical and cultural heritage in a way that is beneficial to both the Park and the surrounding community. The permitted uses of these structures include family homes, artistic and professional activities, rural tourism, restaurants and public facilities. The re-use of these buildings and the associated estate management can also help to preserve the ecological assets of the Park through a public-private partnership (Fig. 9.4).

One important aspect of this proposal is its potential to establish a population with strong ties to the Park, to promote the maintenance of the heritage and to stimulate the regeneration of the area through complementary activities.

9.2.1.5 Infrastructures and Technical Services

The primary goals of PEPNat in this area are to reduce the occupation and fragmentation of open spaces, increase the permeability of infrastructures and establish conditions for the maintenance, improvement or enhancement of existing infrastructures (Fig. 9.6).

In the pre-existing planned areas, PEPNat seeks to adapt planning to the real situation by eliminating non-developed areas and changing the road zoning code for simple paths, among other infrastructure-related proposals.

The Plan also makes a distinction between infrastructures that serve the Natural Park and those that simply pass through it and requires environmental assessment and integration measures for larger-scale infrastructures that cross the space protected by PEPNat.

9.2.1.6 Functional Space

The functional space includes areas outside the Natural Park whose characteristics or functionality enable them to make a decisive contribution to the conservation of the park's values and the processes taking place inside it.

Appropriate organisation and management of these spaces can establish synergies with the Natural Park. PEPNat provides guidelines, recommendations and a catalogue of good practices for this functional space.

These guidelines and recommendations cover topics such as ecological connectivity, ecological values in edge spaces, natural hazards such as fires, accessibility,

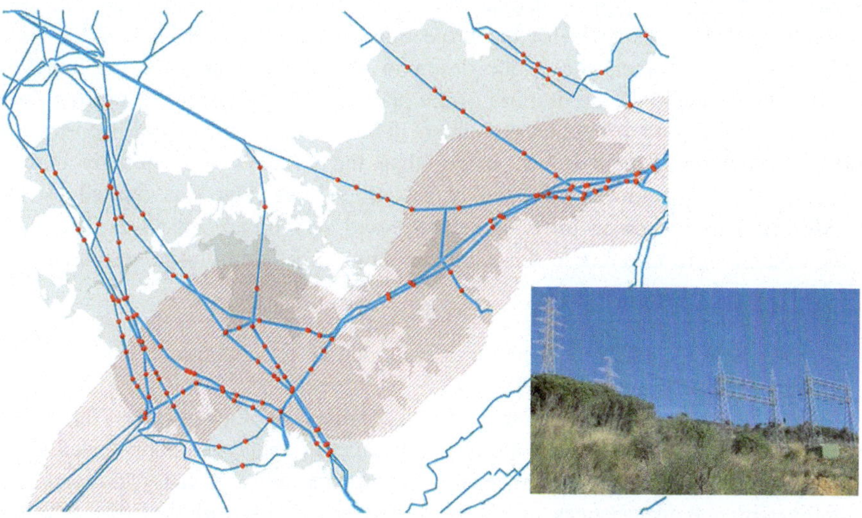

Fig. 9.6 Map identifying problematic electricity lines due to bird collisions. (Source: PEPNat)

Fig. 9.7 Infographic of the entrance to the Park from the functional space in the Finestrelles area. (Source: Batlleiroig)

social use, and the transition between the Park and the urban fabric and buildings located on its perimeter (Fig. 9.7).

The comprehensive approach facilitates the inclusion of a wide variety of key aspects. However, the protection of the ark's landscape and natural environment requires a transdisciplinary approach as well as the systematic collaboration and cooperation between different actors and aspects.

9.2.2 *Systemic Vision*

The creation of the Serra de Collserola Natural Park in 2010 culminated a long process of public demands for the protection of this space. PEPNat is the legal document that channels the public desire to preserve and improve the Serra de Collserola. In addition, the drafting of the plan involved an extensive participation process that included different stakeholders in four phases, with a public information period lasting 6 months (much longer than the 45 days stipulated by current legislation). Furthermore, as stated in section 9.2, PEPNat was drafted by the AMB within a framework of co-governance and very broad consensus (Fig. 9.8). The systemic vision refers to the dialogue between different initiatives and stakeholders on a sustained basis.

The previous section enumerates the most important proposals that illustrate synergies between different thematic areas, such as the functional space and the Technical Estate Management Plan.

The delineation of the functional space is particularly relevant given the Park's central location in the heart of the Barcelona metropolitan area. This space has a major impact on ecological connectivity, the compatibility of ecosystem services, biodiversity and disturbances, public use and the interaction with the urban fabric of the nine municipalities included in the Park.

Fig. 9.8 Main phases in the drafting and approval process of PEPNat. (Source: PEPNat)

With the functional space, the edge ceases to be an administrative boundary and begins to be seen as a strategic space. The Technical Estate Management Plan also summarises the cooperation and synergies between different stakeholders and thematic areas.

9.2.2.1 The Technical Estate Management Plan

One innovative feature that summarises the cooperation of different stakeholders and the synergies between thematic areas is the Technical Estate Management Plan.

As already pointed out, PEPNat aims to promote primary sector activities from an ecological perspective and to rehabilitate the extant cultural heritage (Fig. 9.4). The management of the Park is the responsibility of the CPNSC, a public consortium. However, despite some one-off agreements with other public bodies, the percentage of land managed directly by this consortium is limited. This hinders the overall management of the Park and has a negative impact on the improvement of biodiversity and the promotion of ecosystem services in general and on the promotion of the agroforestry mosaic in particular (Vidal-Casanovas et al. 2021).

The previous urban planning regulation focused on protecting and stopping the transformation/degradation of forested areas. Without abandoning this idea, flexible formulas needed to be found to manage these spaces. Changes were required in the approach, in the goals and in the involvement of all local stakeholders, moving from a position of public control to one of cooperation (Vidal-Casanovas et al. 2020).

PEPNat establishes the obligation to design and have an estate management plan approved in order to develop new uses in existing constructions listed in the Catalogue of *masies*. The management commitments of agroforestry estates will be coordinated with the management body and will be proportional to the characteristics of the estate and the specific characteristics of the uses that will take place there. Local councils may request an economic-financial survey that demonstrates the real possibilities for executing and maintaining the proposed action.

The aspects and measures that need to be included in the management plans are of different types and include both general prescriptions – especially with regard to crops in priority areas for agriculture and the maintenance of paths and field boundaries, and the environmental zoning areas – and the measures detailed on the corresponding page of the Catalogue of *masies*. These mitigation and compensation measures complement the specific proposals and strategic projects to be implemented by the management body in the environmental zones and other key locations in the Park (Vidal-Casanovas et al. 2021).

There are multiple benefits of this co-responsibility formula. They include the maintenance of the agricultural surface and the extensive built heritage, the improvement of biodiversity, the provision of ecosystem services, the mitigation of and adaptation to the effects of climate change, the prevention of natural risks and the supply of local food.

The Technical Estate Management Plan had recently been drafted when the PEPNat was approved 2 years ago. At the time of writing, this concept has been echoed in subsequent key plans and regulations, such as the Metropolitan Urban Master Plan (initially approved on 21 March 2023) and the current legislative framework for urban development issues (DL 1/2010, last updated on 18-03-2023). Both documents include formulas for the co-management of open spaces. The former focuses above all on agricultural and forest land, with the aim of encouraging agricultural activity and forestry management. One of its primary goals is to ensure ecological functionality and to obtain benefits from own resources using the logic of a green and circular economy (AMB SRPD, 2019a). The latter introduces the concept that the use of existing buildings and the custody of affected estates are prerequisites for the authorisation of uses and works in open spaces.

The holistic approach embodied by the comprehensive and systemic vision reinforces the ecological strategy, which is the backbone of the Plan. This ecological strategy is based on preserving biodiversity and enhancing ecosystem services, subject to the evolution of natural capital, through the analysis of the Park's flows and environmental functions. The ecological strategy therefore involves a dynamic and adaptive management.

9.2.3 *Dynamic and Adaptive Management*

The debate on more flexible planning instruments is neither new nor exclusive to open spaces. However, it is in the planning of metropolitan protected areas where this need is most pressing. This is due to three main factors: the dynamism of open spaces, society's demands and different perceptions in relation to the natural environment and the effects of climate change. These three aspects are not antagonistic but are instead interrelated.

Two proposals stand out in this regard: environmental zoning and the monitoring of the state of the Park, with a battery of environmental indicators that are the basis of this monitoring.

9.2.3.1 Environmental Zoning Strategy

In addition to the partial nature reserves defined under Decree 146/2010, PEPNat defines four areas that play a key role in the preservation of biodiversity and the enhancement of ecosystem services. These are the patches of connectivity interest, areas of special importance, islands of tranquillity, priority areas for agriculture and, to a lesser extent, the areas of special significance (see 9.2.1.1 and Fig. 9.3). These areas contribute to the adaptive management of the Park in three ways. First, they have been defined according to their functionality rather than their appearance or degree of 'naturalness'. This approach enables a more dynamic management, by focusing on functional aspects rather than naturalness, which enables work towards specific objectives to be undertaken, with criteria adapted based on their achievement. This does not undermine the regulations established by urban planning laws and PEPNat's own rules for these areas. Second, not only are these areas subject to greater monitoring, but they have also been used to define the first set of indicators. As a result, numerous indicators enable the objectives set for these areas, the status of their ecological health and the actions undertaken within them to be monitored. This facilitates evaluation of their evolution over time, starting when the initial snapshot taken when the PEPNat was drafted. Finally, with the exception of the partial nature reserves, the delimitation of these zones can be more easily adjusted than under the previous plan (PEPCo).

9.2.3.2 Monitoring of the State of the Park

Comprehensive monitoring and surveillance are crucial for obtaining historical data and detecting changes in the Collserola mountain range. This helps in decision-making and adapting the management of the Park according to the proposed model. This monitoring is carried out through the Environmental Surveillance Program [PVA], which addresses both the development of the various measures planned by PEPNat and others that may arise, as well as compliance with the environmental objectives of the Plan and the health status of the Park.

The Park's values and compliance with PEPNat's environmental objectives will be monitored in different phases and scales. Environmental surveillance has been carried out continuously, throughout the strategic environmental assessment and the drafting of the Plan, while monitoring during the implementation phase has been planned at three different levels, based on environmental indicators and the Multicriteria Tool. As a result, the managing body must submit a monitoring report containing information on these three levels to the environmental authority every 5 years.

9.2.3.3 Environmental Indicators

PEPNat has identified 17 indicators that are linked to the environmental objectives. They are grouped into four thematic areas: biodiversity, ecosystem services and landscape, disturbances and education and dissemination. These indicators are applied at two levels: the overall scale of the Park and the management plans for individual properties. While these are the initial set of indicators, they may be updated or expanded based on the information available or specific areas of interest. It is important to note that while the indicators provide information on their own, their significance lies in the interpretation of the data obtained that they all provide together. A GIS-based Multicriteria Tool to monitor the state of the Park has also been planned.

9.2.3.4 Multicriteria Tool (GIS)

In addition to the proposed monitoring, the development of a Multicriteria Tool is also planned, which will be based on geographic information systems, and aims to be a mechanism for monitoring the state of the Park and its values (Fig. 9.9). The objective of this tool is to identify the need to rectify or carry out revisions on an agile, visual and methodologically sound basis, when the evolution of the protected area deviates from the intrinsic park model of the Plan. Whenever possible, the information required to measure the environmental indicators will be obtained through the layers integrated into the Multicriteria Tool (Vidal-Casanovas et al. 2020).

9.3 Final Considerations

There has been a change of paradigm in relation to the conservation of protected areas in recent decades, with a focus on creating multifunctional landscapes that provide ecological, socio-economic and cultural benefits. In metropolitan environments in particular, the role of protected natural areas has shifted from one of resource extraction to multifunctionality, from isolation to connectivity and from exclusion to integration. PEPNat embodies this new approach to protecting the natural

Fig. 9.9 A summary of the most significant interrelated aspects of the Plan to better illustrate this fundamental change in understanding the protection of natural spaces. (Source: PEPNat)

environment and the landscape. Its main objectives are the preservation and improvement of biodiversity and the enhancement of ecosystem services in a dynamic and adaptive management framework. This makes it a plan that is attentive to the Park's values and environmental processes, to the multiple functions and services the protected space offers to the metropolitan territory and population and even beyond and to the interaction with the rest of the green infrastructure.

References

AMB, Servei de Redacció del Pla Director (2019a) Pla Director Urbanístic Metropolità de l'Avanç. Document inicial estratègic. Àrea Metropolitana de Barcelona, Barcelona

AMB, Servei de Redacció del Pla Director (2019b) Pla Director Urbanístic Metropolità. Avanç. Síntesi de l'Avanç. Àrea Metropolitana de Barcelona, Barcelona

Associació Arran de Terra (2017) Diagnosi tècnica: Les experiències de producció agrària del PNSC. Eines per la Dinamització Local Agroecològica. Consorci del Parc Natural de la Serra de Collserola [CPNSC], Barcelona

DL 1/2010, de 3 d'agost, pel qual s'aprova el Text refós de la Llei d'urbanisme. Last updated March 18, 2023, from https://portaljuridic.gencat.cat/eli/es-ct/dlg/2010/08/03/1

Solà-Morales M (2008) Diez lecciones sobre Barcelona: los episodios urbanísticos que han hecho la ciudad moderna/Ten lessons on Barcelona: urbanistic episodes that have made the modern city. Collegi Oficial d'Arquitectes de Catalunya, Barcelona

Vidal-Casanovas E (2022) The Collserola Range Protection Plan [PEPNat]: dynamic and adaptive management of the metropolitan green infrastructure. UrbanNext. Retrieved October 10, 2023, from https://urbannext.net/the-collserola-range-protection-plan/

Vidal-Casanovas E et al (2020) Planificación para una gestión dinámica y adaptativa de la infraestructura verde metropolitana: el PEPNat como caso de estudio. In: Llop C et al (eds) IV Congreso ISUF-H: Metrópolis en recomposición: prospectivas proyectuales en el Siglo XXI: Forma urbis y territorios metropolitanos. DUOT, UPC, Barcelona

Vidal-Casanovas E et al (2021) The Collserola Special Protection Plan (PEPNat): a bid for co-responsibility in agricultural and forest management. In: Catalano C et al (eds) Urban services to ecosystems: green infrastructure benefits from the landscape to the urban scale. Springer, pp 367–387

Chapter 10
Green Regeneration and Transformation Strategies for Climatic Adaptation

Iris Dupper

Abstract More than ever, global climate change comes along with new challenges for the quality of life in urban and rural areas under locally varying specific conditions of our time, which are visible more and more in most people's immediate sphere of influence. The issues surrounding climate protection have reached all parts of society, questioning how our current way of life and economy might affect the climate, or how it might potentially affect society collectively in terms of our everyday lives, the ways we commute and travel or the current price of electricity? This applies in particular to urban space, due to structures defining conditions for a particular urban climate.

Keywords Urban sustainability · Climate adaptation · Upcycling landscape

10.1 Introduction

Climate change is omnipresent in current discussions, and climate zone shifts are already foreseeable, as different prognoses assume a temperature rise of 2–4 C° until the year 2100 (IPCC 2013). Living conditions will change due to the likely increase in extreme events, such as heavy rainfall, storms, heat waves, and droughts. In this context, the anthropogenic influence and the resulting responsibility through the reduction of greenhouse gas emissions or the return to the important significance of undisturbed water cycles play a significant role for the living conditions of future generations, who live with an increasing tendency worldwide in urban areas—according to the World Urbanization Prospects of the United Nations, more than 70% in Europe and North America (Laue 2019).

This article looks into how to deal with leftover spaces under changing climatic factors along examples of urban areas, like in LATZ + PARTNER (L+P Kranzberg Germany) projects in Turin (IT) and Belval (LU), as well as on the fringe of a

I. Dupper (✉)
Partner, Landscape Architect ByAK OAI Lux, LATZ+PARTNER, Kranzberg, Germany
e-mail: i.dupper@latzundpartner.de; http://www.latzundpartner.de

F. Dal Falco (ed.), *Beyond the Garden*, Designing Environments,
https://doi.org/10.1007/978-3-031-50567-6_10

Metropolis, like in Lyon (FR) or Metropolis Rome (IT). In the responsibility to ensure quality of life by applying responsible planning and building measures, the search for strategic usability of intrinsic potentials is essential. The development of municipal green space can offer the chance to favor planning decisions for environmental and societal friendly instruments and planning processes. Therefore, this article focuses on numerous approaches to solutions of this kind in sub-urban public spheres in the central and south European context, which also consider confrontations with elements and factors of local and/or trans-locational climate, also in terms of vegetation. Climatically adapted planning strategies favor the human usability of public spaces and contribute positively to the surrounding mesoclimate. In order to ensure the regulatory capacity of an ecosystem, climate and air environmental improvement are essential components in spatial planning conceptions (Laue 2019).

In addition to climate protection efforts, climate adaptation strategies are also gaining on importance due to the potential of ecosystems of green spaces, which are able to reduce heat stress for animals, plants, and people. Since woody plants are also suffering from climate change, with new associated plant diseases and parasites, the search for resistant climate tree species has begun in the last decade. Tree testing methods in several universities, statal or municipal plant testing gardens or nurseries in central Europe focus on tree's performances in terms of factors like robustness, resistance to heat or drought in urban conditions (Duthweiler 2022). Climate adapted trees species perform well in regard of factors such as the following:

- Carbon storage, as trees absorb and store carbon dioxide, which supports the reduction of global greenhouse gas emissions
- Temperature regulation, since trees can help to lower temperatures in urban areas and reduce heat islands
- Improvement of air quality, since trees can filter pollutants from the air and produce oxygen
- Favoring biodiversity, since trees provide a synergetic habitat for the fauna
- Improve human well-being, as green spaces and trees improve people's well-being, reduce stress and promote physical health (Herzog 2020, based on van den Berg 2012)

10.2 Sustainability as a Key to Urban Prosperity

Public space planning policies are strongly related to political decisions, regarding the management of the encompasses of environmental factors with their effects on "ground" and achievement of "sustainability" as a key to urban prosperity, in relation of the environmental, social, and economic health of cities. While the awareness of the importance of green in urban space seems to be rising, the implementation of measures, like green roofs or green facades on existing architectures or post-industrial architectural elements, seems to offer challenging potentials in vast parts of our cities.

The human experience of outdoor urban space is related to well-being, which also encompasses microclimatic influencing environmental factors with effects, such as the following:

- Radiating cooling, which is strongly dependent on weather conditions, wind direction and surrounding buildings (Stülpnagel 1987), wind effects and turbulence on building edges related to multi-story, and dense and high buildings
- Heat island effects with radiation due to a high degree of sealed urban space
- A lack of evaporative cooling or air exchange effects due to significant weakening of local /regional winds and a lack of open green spaces (Laue 2019) with a minimum of 1 ha in size (Kotremba 2019)

Creating a series of green spaces of at least 1 ha in size, with conditions, which favor the improvement of air quality, oxygen production, air purification, and fine dust binding, is favorable toward few large parks. This is structurally favorable from a microclimatic enhancement point of view, since cooling effects of green spaces of 1 ha in size or more radiate outward between 100 and 300 meters, regardless of the total size of the green space.

In high-density urban neighborhoods or dense residential environments, a series of green spaces of a minimum of 1 ha in size in combination with private green every 200–600 meters show microclimatic consequences avoiding the storage or lack of evaporative cooling or preventing air exchange effects due to low night-time cooling:

- The implementation of green in the urban context, like greenspace by enrichment of biodiversity, retaining rainwater on site, and greening roofs on buildings for cooling by covering and/or evaporation in order to avoid overheating (Herzog 2020 see p. 59; p. 47; on: Choudhry et al. 2015; Houlden et al. 2019), shows enhancing effects on a microclimatic scale.
- Integrated green facades or green facades in front of façade openings show enhancing effects on human well-being, like health benefits of contact with nature or the principle of the feedback loop of aesthetic driven human environment interactions taking into account human perceptual processes (Herzog 2020, p. 23, 47; Gobster et al. 2007, p. 963), due to the strong visual presence of plants and their natural cycles, which can influence the user regarding principles.

Sustainability is the key to urban prosperity. Planet-friendly foundations and strong local authorities supporting the vibrancy of their communities and shared economic value seem to be essential pillars for prosperous, sustainable communities. Since there is a tendency for development solutions focusing on sustainable city principles, where ground and open space is used diligently and with a social and sharing sense, this leads to an increase of the role of sub-urban peripheral spaces, connected to new hybrid forms of living environments. Changes often take place more on a local, than on a national level, bringing up new emerging parameters of urban development. According to these new parameters, nationwide three-quarters of the population live in built-up or heavily built-up agglomerations (Destatis 2019) and with a high percentage of young adults and single households (Heimerl et al.

2019). This percentage influences growing demands for education, mobile networks, and micro-societies. Another challenge is the necessity to shift toward nonfossil energy systems and a more climate-adaptive systems, shifting regard from a mobility car-based city model into a more suitable human scale city. This transition requires more methods and instruments for improved coordination of an integrated social, urban, and infrastructure development.

10.3 Selection of LATZ+PARTNER Projects Focusing of Green Regeneration and Climate Adaptation

The following insights in L+P project work show a selection of aspects in mostly current projects integration the topic of green regeneration and climate adaptation along the mentioned parameters. These approaches demonstrate the research for rules how to transform a site by using intrinsic potentials, like what existing elements can be re-used for the construction of a new park or which elements need to be dismantled as quickly as possible, which pollutes soil elements need to be removed from site as quickly as possible, or which uses can be invented in order to re-use existing elements for the new park, conserving them at the same time authentically for shaping a new identity?

These L+P examples have characteristics in common, showing how architectural leftovers of postindustrial elements can be positively transformed into new landscapes with diverse fields of action by landscape architectural strategies daring to explore potentials of re-use or recycling/upcycling of existing elements, combined with vegetative milieus understood as a common habitat for species of flora and fauna in the dynamic synergy process of increasing biodiversity rate. Resilient green measures serve as a basis for strategic transformations, enhancing social landscapes and microclimatic conditions for regenerative spaces. Leftover areas in the peripheral nodes inside and outside the city center, where public space might initially appear as a network of spatial "voids," can offer potential for public urban spaces characterized by connecting the urban with the suburban fabric.

L+P projects, like the parking SHEDs next to the Esplanade TASE, Métropole de Lyon (Textile Artificiel du Sud-Est, 2016–2021), research the public urban and suburban fabric along criteria such as green integrated in multi-coded open structures and re-used, recycle, or upcycled architectural elements for local uses and regeneration. These new landscapes are open for flexible use and spontaneous actions and appear as milieus of high biodiversity, social well-being, and inclusion.

Studying realized L+P transformations in public space in different scales, along the projects Parco Dora Spina Turin (2004–2012) and Sinter basin square Belval (2018 to today), the combination of social and cultural parameters with influencing factors can be identified, like shaping new identity, regaining a sense of belonging. Re-activated architectures and circularity formulate by climatic adaptation with new green an attitude towards a new aesthetic for action and urban sustainability.

The environmental vision "Regenerative Tiburtine Millieu" (2009 to today on nonpublic ground) researches cultural and territorial strategies of actions toward a public parc for the Eastern Roman Metropolis and the city of Tivoli.

L+P project experiences explore these strategies along green regeneration and transformation strategies for re-activated architectures and postindustrial leftovers since many years in the office L+P, which was founded by Peter and Anneliese Latz 1968 in Aachen, Germany. Tilman Latz explores, for example, within the Parco Dora Spina project topics, like methods of re-use of existing postindustrial structures for multifunctional actions offering synergies already by their pure material value. In the Esplanade TASE and Sinter basin, square Belval project Tilman Latz and Iris Dupper investigate the potential of resilient green milieu in new public urban space over three levels integrating re-activated postindustrial leftovers.

Iris Dupper researches in "Regenerative Tiburtine Millieu + Route" (2009 to today) the potential for a strategy of actions toward an urban resilient parc and biosphere reserve for the Eastern Metropolis Rome, where the quality of the natural environment in a city is ranked higher (Rome ranks 34th out of 100 major international cities), than the urban and social perspective (Rome ranks 68th out of 100 major international cities) according to the Sustainable Cities Index 2022 (Arcadis–Citizen Centric Cities 2022).

10.3.1 Parking SHEDS Usines TASE, Carré de Soie, Vaulx-en-Velin and Villeurbanne, Grand Lyon Mètropole

The long-term Carré de Soie project was implemented since 2007 as a metamorphosis of a vast territory. The built results use the potential of this major metropolitan site and connect to public transport systems, for the benefit of all. Impelled by a citizen's initiative, the design was carried out via requests and objectives in a residents' consultation process. Principal project goal was to keep local silk production history alive and to create a green oasis in a densely built urban neighborhood. Public space inserts local heritage in a respectful creative way and invite for social communication and enhance inclusion. The rusty red color of the shed walls and new walkways are powerful elements and traces. They create a new identity, orientation, and legibility of this space, whose industrial character is even strengthened by the striking aesthetics of upcycling.

The development of the parking SHEDS integrates upcycling processes of transformed formerly indoor concrete floors of historic demolished machine halls of TASE and new green.

The parking project was planned and realized in the context of the Esplanade TASE in Vaulx-en-Velin, designed by L+P, on the site of the demolished machine halls of the heritage silk production site founded in 1925. The parking, for approximately 40 cars and bicycles, is reserved for users of the preserved sheds, which provide space for new companies and workshops as part of a program of the Lyon Mètropole. The creation of a 3000 m2 parking with a contemporary innovative character, respecting and highlighting equally the industrial character, was the aim (Fig. 10.1).

Fig. 10.1 Upcycling process of concrete surfaces in fertile green. (Source: L+P 2021)

In order to minimize resource consumption, existing materials, as the original former inner concrete flooring surface, were largely reused for the new parking: In particular, the concrete floor plate of the former halls served as the base of the parking lot with planting area, cut out of the concrete.

The concrete rubble was sorted by grain size and re-brought in as a mixture of crushed stones, earth, and stone or mulch (depth – 0.6 m + drainage layer −0.5 m). The cutout surfaces were filled with substrate for new vegetation (depth −1.5 m + drainage layer −0.5 m) consisting of species of rockery plants, ornamental grasses, perennials, shrubs, and trees of small and large size, which will gradually develop into shady green ecosystem in this barren landscape.

The striking orange color (RAL 2005 and 3026) of the original indoor shed walls and concrete surfaces, with inscriptions and traces of the industrial past, is now powerful outdoor reused elements of the new project. The key elements are duplicated and developed further for the parking lot design: An orange corridor marks the pedestrian connection between the vertical wall and the parking lot surface, and the interplay of lettering and graphics serves as signaling. Colors and signs lend identity and legibility to the space, whose industrial character is authentically strengthened by the striking aesthetics of upcycling processes (Fig. 10.2).

Fig. 10.2 Applied upcycling principles for transformation of concrete slab into a sustainable parking. (Source: L+P 2023)

Parking SHEDs planning and realization: 2020–2021.
Client: Métropole de Lyon.
Landscape Design, Urban Planning: Latz+Partner (Mandataire, lead design Tilman Latz, Iris Dupper, project lead Laura Ruccolo).
with MRP Paysage and TECTA Engineers.

10.3.2 Les Usines TASE (Textile Artificiel du Sud-Est, 1935–1980) in Carré de Soie, Vaulx-en-Velin and Villeurbanne, Grand Lyon Mètropole, France

In 2004, Greater Lyon appointed Bruno Dumetier (AABD) as chief urbanist for the urban project "Carré de Soie" covering 500 ha in size with surfaces, where heritage was taken as one of the pillars for the future key development of the area in Eastern Lyon. The Esplanande TASE was appointed as pilot project of the Mètropole de Lyon, Vaulx-en-Velin, and Villeurbanne. The urban project "Carré de Soie" integrated the project Esplanade TASE as pillar for the development of the area based on Lyonnais industrial heritage and as one of the former 14 silk productions plants in Eastern Lyon.

The silk production plant was built since the mid-1920s by architect Desseux in Vaulx-en-Velin, as one of the first sites for the industrialization of artificial silk (viscose) in France. In 1935, the company TASE, employed up to 3000 people and then end of the 1960s around 1500 for the production and invention of new fabrics in "artificial silk" in France. In 1980, production stopped, and the site was closed. The long main factory façade to the south overlooking an esplanade is an example of Tony Garnier-style architecture.

Based on upcycling processes, the concrete floor surfaces of the former architectural workshop sheds were transformed into new open public green structures. The process was supported by the expectations Métropole de Lyon combined with the implementation of an initial participation process with discussion tools consulting future residents on their expectations toward a green Esplanade as of public space. Inspired by a citizens' initiative, areas were not built over, but developed into public open space and a park of high quality, also in terms of biodiversity, partly planted by joint participatory campaigns with citizens. This led to a dialog in participative workshops, balancing the components "understanding" and "creativity," to eco-educational actions with pupils of Odette Cartailhac School and the realization of the communal gardens, Potager-en-soie, which are 100% managed by residents.

A principal goal was to keep local history alive and to create a green oasis in the midst of densely built neighborhoods. The existing water tower and the listed factory façade bear witness to the industrial past and become integral elements of the new open space. Reused postindustrial surfaces and upcycled emblematic architectural elements, define today one of the flagship public spaces of the Carré de Soie since summer 2021 the TASE esplanade, near the water tower, offer a new breathing space for inhabitants, employees, and visitors (Fig. 10.3).

Fig. 10.3 Section through upcycled concrete soil surface into a green promenade. (Source: L+P 2016)

This combination of existing, partly listed architectural elements, set in a new urban context, by means of contemporary landscape architectural and planting interventions processes, evoked new aesthetical qualities, public spaces for communication, and newly initiated social processes of inclusion with new inhabitants and youth activities. These measures have given inhabitants, employees, and visitors a new sense of belonging to the site. The popularity of this project is also based on an emotional connection to this site with highly frequented publics spaces, used by local users of all ages, being able to stay in their local neighborhood, and well connected to public transport and cycle lanes, avoiding individual car traffic. These impacts and results show how this approach can be exemplary and in line of the latest building technics in relation to sense of belonging (Fig. 10.4).

For the team around, L+P heritage was a key pillar for the transformation of this postindustrial site into the public green Esplanade TASE. Existing elements were reused and transformed, like the golden passageway. Designed as a multipurpose area, the esplanade is a green space for people of all ages and an area of social inclusion enabling positive intergenerational exchange and community building. More than 260 trees were planted (until August 2020), to shape the recreation area nearby a future school, with play and communication areas, as a place to picnic and relax in deckchairs or in intimate golden niches in the rammed earth concrete wall to shelter the park from infrastructural lines (Fig. 10.5).

The unsealing of the car park equally created new identity, as the rammed earth concrete wall, evoking the new spirit to place, which is also marked by the integration of the significant historic water tower.

By transformation strategies, the new green Esplanade has become a binding factor for the community, a positive place for communicational exchange between

Fig. 10.4 Planted promenade. (Source: L+P 2023)

generations—users knowing the project, before it was finished. New aesthetics evoke a dialogue with the existing and new experiences. Sustainability was seen here as an asset and not a constraint offering 2.8 ha of new public space for regeneration in a green surrounding considering the climate adaptation measures in reactivated industrial architectures, without significant misuses up to now.

Esplanade TASE planning and realization: 2016–2021.

Client: Métropole de Lyon - pilot project of Mètropole de Lyon, Vaulx-en-Velin and Villeurbanne.

Landscape Design, Urban Planning: Latz+Partner (Mandataire, lead design Tilman Latz, Iris Dupper, project lead Laura Ruccolo, Sophie Holzer) with MRP Paysage and TECTA Engineers, EGIS, Marc Aurel et Arènes / Lyon.

Awards: Yuanye Award 2021 category Urban Design in Silver.

10.3.3 Sinterbecken Belval, Luxembourg

In 2001, a national investment program was set up to transform the conversion site in Belval into an urban district by the masterplan for Jo Coenen. The common grounds of Esch-sur-Alzette and Sanem correspond to the size of 170 football fields.

In Belval-West, the ArcelorMittal steelworks has ceased operations in 1998. The company started working together with Luxembourg State on initial studies and considerations that ultimately lead to the development of a masterplan for the future Belval district, the concept of the "Cité des Sciences, de la Recherche et de l'Innovation."

Fig. 10.5 Rammed earth concrete wall with golden niches. (Source: L+P 2020)

Luxembourg was defining to convert the patrimonial industrial wastelands into reusable land again, which in the case of Belval was entrusted to the newly founded development company AGORA in October 2000. AGORA called for a final master plan determining the urban planning concept and landscape design of the new quarter "Cité des Sciences," as a statal key project. The new concept also included the two blast furnaces, adding to shape the Belval's silhouette.

In 2002, Jo Coenen & Co office's Masterplan included specifications like sustainable construction, renewable energies, and a mobility concept, defining that the design of the park and the entire landscape should be considered exemplary.

In 2008, AGORA adapted the master plan to the new real estate market conditions, by reducing the lots in the Square Mile quarter. In 2010, the emphasis in the

Esplanade area focused on office space for small and large companies. In 2011, the Square Mile sub-quarter was redesigned by KCAP.

The public square around the patrimonial old Sintering basins was decided by an international competition, won by Metaform Architects (mandatory) with L+P (landscape architects) and HLG (engineers). The last patrimonial element is to be transformed into a sustainable green public square, as new pulsating heart of the urban Central Square quarter.

7000 m^2 of new public space on three levels transform the currently inaccessible and mineral holes, into a new public space. Connected by green, regeneration strategies manage to preserve the existing levels -1 and -2 in combination with the newly raised square level. Climate adapted tree species together with a vibrant vegetative milieu over 8 height meters manage to preserve the industrial heritage of national importance, make it accessible and usable, as well as to contribute in a regulatory way to urban micro-climate adaptation. Species of nature shall reassert itself in this area, which seemed to be too minerally transformed and "lost for Nature forever" (Fig. 10.6).

Reuse and upcycling of architectural elements form the structural base for vegetation in the lower levels incorporating fertile soils for vibrant green in a unique setting around a circular water pond on -2 level overlapped, by a sunken suspended deli/café (Fig. 10.7).

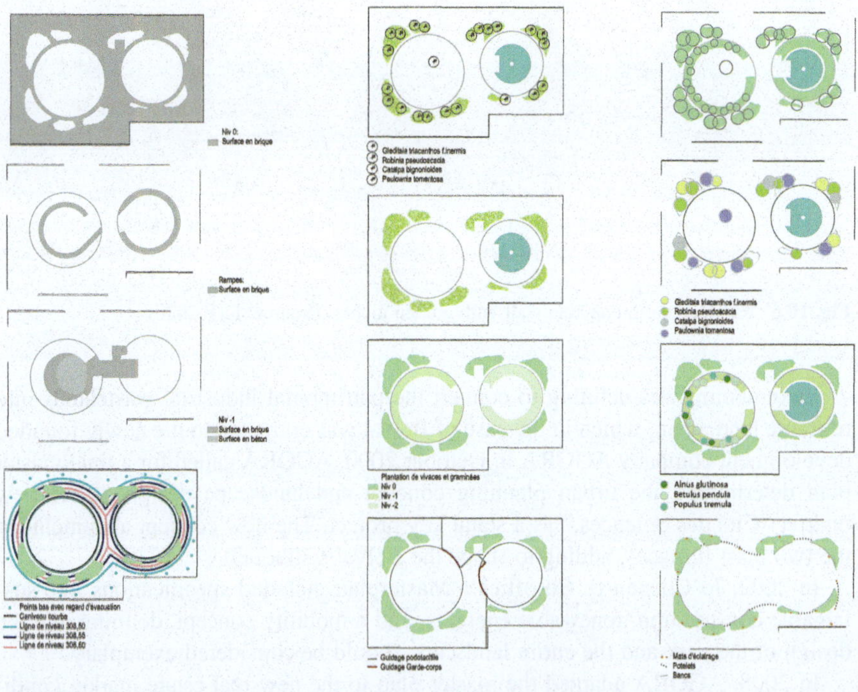

Fig. 10.6 Vegetative milieu over three levels. (Source: L+P 2022)

Fig. 10.7 Reuse of industrial elements for public Sinter basins square. (Source: L+P 2022; site image source: Jessi Lang 2021)

The sustainable green strategy of the vegetative milieu over three levels connects as a stepstone toward the unique Minett Unesco Biosphere Reserve on the territory of Belval and the bordering Southern Region, characterized by its postindustrial leftovers.

Belval and the future Sinterbeckenplace are situated in the heart of this new biosphere reserve.

When industry ceased its activity in the region toward the end of the 1990s, nature reclaimed its terrain. Unexpected and rare flower species, like orchids, appeared on the rough grasslands and wet meadows, accompanied by a very special fauna, which had settled here, like rare butterflies, bats, amphibians, lizards, reptiles, and birds.

As the Minett Unesco Biosphere project is also about showcasing the legacy of the industrial age as it appears today amidst biodiversity and a new sustainable urban neighborhood, Belval also has a special appreciation for its modern architecture, which has been artfully integrated into the industrial heritage of the former Arcelor-Mittal production sites.

Sinterbeckenplatz, Belval Sanem (Competition 2018–APS 2022 to today).
Client AGORA s.à r.l. et cie, secs.

Landscape Design, Urban Planning – Latz + Partner (competition and lead design Tilman Latz, Iris Dupper; project lead APS Laura Ruccolo) with Metaform Architects Luxembourg (mandataire), HLG Engineers Luxembourg.
Awards: WAFX Winner 2023 Category: 'Re-use'.

10.3.4 Regenerative Tiburtine Milieu + Route – Regional Vision

The idea for a biosphere reserve on the currently nonpublic and inaccessible territory of the former travertine quarry leftovers and transhumance route of the Western Tiburtine Region originated in 2006, in the context of my research in relation to the Rome prize-Villa Massimo scholarship.

The regional vision Tiburtine Milieu and Route are aiming for long-term action for over 350 ha, in combination with patrimonial cultural transhumance pathways crossing sulfurous springs in the region Tivoli and Eastern Roman Metropolis. Only few biosphere reserves offer special nature on formerly industrially quarried heritage sites since Roman ages, for example, 78 ha Barco with authentic travertine quarrying traces of Lapis tiburtinus found here.

Since 2009, research key goals were set, highlighting the importance of newly developing natural biodiversity, developed on quarrying waste hills, like Montarozzo del Barco qualified as prototype, for the protection and development of the ecosystem.

Innovative approaches to sustainability can be potentially found in economic development, for example, the use of water power, implementation and use of renewable energies on recycled building elements, or the implementation of quiet zones by land development, for the protection of natural, cultural, and industrial heritage, environmental education, and enhancement of the relationship between human and nature.

Today, for example, there are many postindustrial ruins on the western slope of Tivoli, which were formerly installed to exploit the hydroelectric potential below the Acropolis. Available energy is the prerequisite for any value creation. Since in Tivoli water, historical knowledge about the production and use of clean hydropower, as well as architectural and postindustrial ruins of several centuries, are available for reuse or upcycling processes. Sustainability in a cultural setting and biosphere reserve could become useful for this region by strengthening intrinsic potentials for resilience by green regeneration and transformation strategies for climate adaptation.

Tivoli with its Aniene waterfalls and the Sanctuary of Ercole Vincitore/Via Tecta are situated along the authentic Tiburtine Route, connecting directly to the heart of the future envisioned regional park leading 120 m height meters down to the West, to the new biosphere reserve (Fig. 10.8).

Fig. 10.8 Tiburtine Route and Milieu. (Source: Iris Dupper (2015), Montarrazzo Giardini (2013), p. 165 (2))

Regional vision – public parc on the grounds of Eastern Metropolis Rome and the Municipalities Tivoli, Villanova and Guidonia-Montecelio.

Regenerative Tiburtine milieu + route.

2009 to today.

Iris Dupper.

Presented at 14. ICOMOS Conference Florence, Italy 2014; 52. IFLA Congress St. Petersbourg, Russia 2015; ECLAS 2020 Sweden "Stop and think."

10.3.5 Parco Dora Spina Torino

Turin industrialization began at the end of the nineteenth century along the banks of the river Dora. It reached its zenith in the early twentieth century with the construction of the Fiat Ferriere Piemontesi steel, and sheet metal works and the Michelin tyre factory, which shut down in the general decline of the industry during the 1980s. The post-industrial leftovers were a large area of urban dereliction in the city center, in addition to many small pockets of disused land.

With the urban renewal program (*Programma di Riqualificazione Urbana PRIU*) in 1998, these areas of postindustrial dereliction were regenerated with new uses. From Turin's largest intra-urban industrial wasteland emerged a 37 ha new urban park. New interventions line up along a development axis, called "spina" (backbone). With 45 hectares, "Spina 3" is the largest project within the comprehensive

Fig. 10.9 Green regeneration in Parco Dora Spina. (Source: Ornella Orlandini 2016)

structural redevelopment. Due to the reuse of industrial heritage and upcycling of architectural elements in open structures in the Parco Dora conception, a new understanding of inner urban landscapes with new aesthetical qualities emerged, which equally reflects societies' transitions.

The park comprises five separate areas ("Lotti"), three of which are named after the industrial companies that used to occupy the site: Ingest, Vitali, and Michelin, in addition to Valdocco North and Corso Mortara, situated above the new tunnel. The site is dominated by the river Dora, main traffic arteries and new residential areas, but its true and unique character is derived from the reused of its postindustrial elements. The inner connections between the five areas of the park's and the connections with the surrounding neighborhoods were an essential element in the urban sustainability concept for the new park (Fig. 10.9).

The aim was to incorporate the individual character of each area, to strengthen, and to enhance it with new elements, in order to create a unified park experience with reused architectural elements in a 37 ha new urban park with public spaces, transformed by green regeneration strategies and climate adapted spaces for reactivated fields of action (Fig. 10.10).

Parco Dora Spina, Italy 2004–2012.
Client: Commune di Torino.
Latz + Partner (Mandataire - lead design Tilman Latz, Peter Latz; project lead Dörte Dannemann, Daniela Strasinsky); STS S.p.A., Bologna, CMC, Turin, Studio Pession Associato, Turin, U. Marano.

Fig. 10.10 Parco Dora Spina hosting new fields of action Kappa Futur Festival 2022. (Source: MovementMultimedia Sinestésia)

Kappa future music Festival from 01.-03.07.2022 takes place in Parco Dora Spina Torino.

Awards: European garden prize 2017 (EGHN, 2nd prize); International Architecture Award 2012 (Chicago Athenaeum); Premio Architectture Rivelate 2012 (Ordine Architetti de Torino& Fondazione OAT).

10.4 Conclusion

These examples demonstrate that a call for an intensified examination of the social potential in combination with the current urban climate and the associated planning discussion is necessary. Urban and suburban open spaces offer great potential for direct mitigation for social public uses in combination with the adaptation to changing climate conditions. For urban climatic structures, also with regard to habitats of flora and fauna, a fast adaptation to rapidly changing climatic conditions seems essential.

The overuse of fossil fuel-based building substances in urban space, like cement (Heringer et al. 2002) or asphalt, comes along with environmental and societal consequences, which is why the reduction of sealed surfaces and augmentation of useable green surfaces, which serve the public, reduce heat islands by fresh and cold air production supply, add insights into contemporary debates, like strategic social soil management for climate adaptation (Kotremba 2019; Laue 2019; Alberti 2019; Garau et al. 2015).

References

Alberti V (2019) The future of cities. 1st edn Publications Office of the European Union. Retrieved 18 September 2019, from. https://ec.europa.eu/jrc/en/publication/future-cities

Arcadis–Citizen Centric Cities (2022) Europe Sustainable Cities Index 2022″. [online] Retrieved 2 July 2023, from https://www.arcadis.com/en/europe/our-perspectives/sustainable-cities-index-2022/europe/

Destatis (2019 & 2023). Statistisches Jahrbuch Deutschland und Internationales.. Retrieved 3 September 2019 from https://www.destatis.de/DE/Themen/Querschnitt/Jahrbuch/statistisches-jahrbuch-2019-dl.pdf?__blob=publicationFile. https://www.destatis.de/DE/Im-Fokus/Klima/_inhalt.html

Duthweiler S (2022) Lecture ‚Pflanzenverwendung im Wohnumfeld. Stadt- und Strassenbäume im Klimawandel. BYAK Nr. 22122‘. (Based on: GALK Strassenbaumliste; Feldversuche bayerische Landesanstalt für Wein- und Gartenbau Veitshöchheim; KlimamatrixArten Prof. Dr. Andreas Roloff Uni Tharandt)

Garau P, et al (2015) Global public space toolkit: from global principles to local policies and practice. United Nations Human Settlements Programme. Retrieved 25 September 2019 from https://unhabitat.org/wp-content/uploads/2015/10/Global%20Public%20Space%20Toolkit.pdf

Giardini M (2013) La flora vasculare del Montarozzo del Barco. http://www.museocivico.rovereto.tn.it/UploadDocs/4972_art08_giardini

Heimerl A, Folda H, et al (2019) Zukunftsschau München 2040+ Szenario-Prozess und Werkstattreihe. Stadtentwicklungsplanung Wohnen, Bevölkerung und Perspektive München. München: Landeshauptstadt München Referat für Stadtplanung und Bauordnung.. Retrieved 5 October 2019 from https://www.muenchen.de/rathaus/dam/jcr:d46d2325-619e-4c46-97f8-5883d92c4b10/LHM_Zukunftsschau_Web_01-1.pdf

Heringer A, Blair Howe L, Rauch M (2002) Upscaling Earth – Material Process Catalyst. GTA Verlag ETH Zürich. p 23

Herzog L (2020) Green and Shadows. PhD Planning, Design, Technology of Architecture. La Sapienza Roma. Chapter 1.2; p. 23; p. 25 based on van den Berg at al. 2012, p. 53; Fig. 19, p. 47 based on: Gobster et al. 2007, p. 963; p. 59 based on: Choudhry et al. 2015 and Houlden et al. 2019. pp 277–280

IPCC (2013) In: Laue 2019, p. 12

Kotremba C (2019) Kommunales Grünflächenmanagement als wichtiges Steuerungselement einer klimagerechten Stadtentwicklung und Förderung von Biodiversität. Ruelzheim. pp. 6–7

Laue H (2019) Klimagerechte Landschaftsarchitektur. Patzer Verlag Berlin. p 12; 54–57; 155; 161

Stülpnagel v (1987) In: Kotremba 2019. pp 6–7

Chapter 11
Experience an Integrated Approach: Combining Sustainability and Aesthetic Quality with Inclusiveness and Equity Through a Multi-scalar and Multidisciplinary Design of Green Systems

Maria Beatrice Andreucci, Irene Poli, and Chiara Ravagnan

Abstract This concluding chapter outlines key research paradigms that influence the way we understand the benefits of nature and how new directions of integrated approaches – multi-disciplinary and multi-scalar – can innovate regenerative design and policy. The renewed attention to the benefits of green blue infrastructure for both human and climate health was pointed out by all the contributions during the study day. This came because of the growing awareness, in the post-COVID-19 era, of the role that nature plays in human health and well-being, coupled with emerging social economic issues and environmental degradation, especially in the urban built environment. In the first part, the authors point out how biophilic design theory sits in this field, facing sustainability challenges, and how it can be and has been applied at different scales in the urban built environment, mainly referring to the architectural scale. In the second part, the authors pay attention to the role of green systems in the reconfiguration of urban morphology for the aesthetic quality of cities, with reference to the urban fabrics scale. In the last part, the authors highlight how green networks can contribute to the strengthening of safe and inclusive social interactions, as well as eco-friendly lifestyles, with reference to the urban and territorial scale, underlining, in this way, the transversal character of the approach identified in the context of the reflections carried out in the book, which also includes disciplinary contributions traditionally sectoral, and pursuing the objective of innovating the archetypal concept of 'garden', identifying the city as an Open Laboratory.

M. B. Andreucci (✉) · I. Poli
Department of Planning, Design, and Technology of Architecture, Sapienza University of Rome, Rome, Italy
e-mail: mbeatrice.andreucci@uniroma1.it; irene.poli@uniroma1.it

C. Ravagnan
INU, Rome, Italy

F. Dal Falco (ed.), *Beyond the Garden*, Designing Environments,
https://doi.org/10.1007/978-3-031-50567-6_11

Keywords Biophilic design · Urban morphology · Green infrastructure · Green systems

11.1 Introduction

There is a heightened awareness of the role that nature plays in human health, urban environment and social practices. In particular, the Covid-19 pandemic that has coincided with pressing climate and social concerns has strengthened the attention on the potential of natural features in terms of climate adaptation, aesthetic quality and local identity. In this context, the most advanced fronts of debate and experimentation in Europe show the emergence of new methodological and operational references and practices that place the construction of green networks at the centre of urban and territorial regeneration processes as a response to the vulnerability of physical and social structures and to situations of degradation and abandonment of increasingly significant and strategic parts of the territory, internal and external to urban systems. This debate pays particular attention to the benefits of green systems for both human and climate health, as pointed out by all the contributions. In this framework, in consistency with the reflections developed in the two different sections of the book, this conclusion chapter outlines key research paradigms that influence the way we understand the benefits of nature and how new directions of integrated approaches – multidisciplinary and multi-scalar – can innovate regenerative design and policy. Firstly, the paper explores how the *biophilic design* approach and theory fits into the field of sustainability challenges and how it can be and has been applied at different scales in the urban built environment, with particular reference to the architectural scale. The paper then considers the role that *green systems* could play in reconfiguring urban morphology for the aesthetic quality of cities, with particular reference to the scale of the urban fabric. The reflection then focuses on how green networks can contribute to the strengthening of safe, equitable and inclusive social interactions, as well as environmentally friendly and sustainable lifestyles, with particular reference to the urban and territorial scale. The paper underlines the transversal nature of the approach identified in the context of the reflections carried out in the book, which includes disciplinary contributions that are traditionally sectoral, such as psychology, ecology and law, and which aim to innovate the archetypal concept of the 'garden' in theoretical, methodological and operational terms, identifying the city as an *open laboratory*.

11.2 Biophilic Design for People and the Environment

Global health emergencies have emphasised the multiple benefits of nature in buildings and open spaces for social, physical and mental health. Salutogenic research suggests that our affiliation with nature, and in particular biophilic design, may be key for improving health, well-being and quality of life (Antonovsky 1987; Kellert and Wilson 1993; Andreucci et al. 2021).

This renewed attention is supported by an inclination in architecture and urban design that is trying to provide prospects to connect people with nature through regenerative and biophilic design projects and interventions, which have been related to increased well-being, focus, socialisation, sense of place and relationship with nature (Beatley 2016).

The most influential research programs in the last century have been based on adaptive or utility paradigms. Evolutionary psychologists presume all human behaviours reflect the influence of physical and psychological predispositions that helped human ancestors survive and reproduce.

There are two research programs that have emerged out of an adaptive paradigm that have gained the utmost attention and subsequent experimental practice. The first is centred on restorative settings that help with the restoration of attention or to improve perception and awareness, notably Stephen and Rachel Kaplan's Attention Restoration Theory (ART) (Kaplan and Kaplan 1989; Kaplan 1995). The second emphasises the ability of healing environments to encourage stress recovery and positive mood, notably Roger Ulrich's Psychophysiological Stress Reduction Theory (PSR) (Ulrich 1983).

In their influential research on restorative environments, Kaplan and Kaplan (1989) have pointed out the importance of four restorative characteristics of nature in promoting positive effects, namely, being-away, referring to a change of scenery and/or experience from daily routines, promoting a conceptual distance from the ordinary; fascination, intended as the capability of nature to involuntarily elicit the individual's attention, without mental effort and thus the depletion of cognitive resources; extent, implying the properties of coherence among the environmental elements and scope in environments, which should be perceived as extended enough to engage the mind; and compatibility, which has to do with the perceived congruence between the characteristics of the environment and people's needs, intentions and inclinations. This theory has been extensively verified in later studies (Kaplan and Kaplan 1989; Kaplan and Kaplan 2005; Hartig et al. 1991). A key constituent of investigations assessing ART has looked at aesthetic human inclinations towards different kinds of nature. These findings argue that some forms of nature are more promising to restoration than others and that natural sceneries are more revitalising than urban settings (Bratman et al. 2019; Kaplan and Kaplan 2005; Korpela et al. 2008).

Studies challenging the Psychophysiological Stress Reduction Theory also adopt an evolutionary biology theory but tend to give emphasis to the emotional or sentimental aspects of this relationship. At its core, evolutionary biology argues that because we evolved in nature, we tend to feel connected with things that remind us of nature; this attitude is called biophilia (translated as a love of nature) (Ulrich 1993). This affiliation with nature has begun to be considered for its potential to link to biophilia, which has been revealed to support human health and well-being as well as more sustainable behaviours and acceptance of urgency to fight climate change (Beery et al. 2015). While the utility paradigm also draws on the idea that our natural environment relates to our well-being, it focuses on the role that nature plays as a value of a place to satisfy interpersonal needs of affection, control and belonging as interdependent and variable. These are often measured by known benefits of access to nature, such as increased levels of physical activity, healing experiences, social acceptance, interaction and shelter (Braubach et al. 2017).

With the growing research and implementation dedicated to biophilic design, it is interesting to look at buildings specifically designed and constructed as archetypes to emphasise the biophilic indoor qualities.

With specific reference to the design works presented at the Conference 'Beyond the Garden', a general result emerges from our analysis showing that natural environments can promote higher restoration also when compared to artefacts with similar restorative potential. This consideration supports the existing empirical evidence that nature as such – and not merely because of its restorative characteristics – is healing and revitalising, in line with ART (Kaplan and Kaplan 1989; Kaplan 1995), SRT (Ulrich 1983) and, overall, the Biophilia hypothesis (Kellert and Wilson 1993).

11.3 Green Systems to Reconfigure the Urban Morphology

The issue of green networks is a disciplinary field that has been widely addressed in the debate of the last decades, especially in Italy (Secchi 1989; Macchi Cassia 1991; Gambino 2007; Gasparrini 2015). However, the consideration of green networks as structural elements of the urban form takes on particular importance in the face of the growing awareness of the consequences of the fragmentation and trivialisation of public spaces in contemporary sprawling cities, as well as the criticisms associated with the absence of green spaces in the historic and consolidated urban fabric. The fragile contemporary urban context is subject to the impoverishment of ecological resources, the increasing frequency of extreme weather events and the degradation of urban landscapes, which are the custodians of place identity, social cohesion and inclusiveness (Borja and Muxi 2003) and the quality of the *cadre de vie* (Gasparrini 2015).

Starting from the awareness of the deep relationship between physical continuity and social inclusion and between the quality of networks and the quality of the city (Gehl 2010), current urban planning and design strategies face the reconfiguration and strengthening of green networks as guidelines for a resilient city (Marcus and Colding 2014), open to global and contemporary challenges and sensitive to historical and local roots. Green networks aim to rebuild relationships, thus acting as a reference both for the construction of an ecologically oriented public city, for the reorganisation of the economic base and the triggering of a new urban metabolism (Gasparrini 2015) and for the reduction of vulnerability to environmental risks and land consumption. At the same time, an interconnected and multifunctional environmental system can play a fundamental role in recomposing and strengthening the urban morphology, as a strategic tool for mending the physical and social fragmentation of the contemporary city. A branched and complex connective that links and interacts open spaces and anthropic contexts, *going out* towards the extra-urban areas of settlement dispersion and *coming in* to reach the meshes of the compact city. A connective that is capable of reconfiguring the places of abandonment and decommissioning, contrasting marginality, anchoring itself to the systemic and identity character of common goods (Ravagnan and Poli 2017). And conceived in strongly inter-scalar terms as a structuring choice of invariants and a reference framework for urban planning and design at large scale, as a tool for managing sustainability at local and municipal scale, and as a driver of the quality of interventions at the scale of each single transformation (Masboungi 2012). Faced proactively with an integrated strategy of urban regeneration, reconnection of natural and rural areas, environmental rebalancing of natural features at urban level and rehabilitation of abandoned sites, mending the dense historic and consolidate urban part with the fragmented settlements, including in a unique framework the network of public spaces, hydrographic networks, historic parks, agricultural systems, infrastructure networks. Green networks also combine this grid of morphological and environmental references for the sustainable and resilient regeneration of the territory, with public, semi-private and private spaces within compact fabrics, such as through Christian de Portzamparc's concept of the 'îlot ouvert' (Semapa 2010). A morpho-typology used, for example, in the French *ÉcoQuartiers*, is characterised by the quality of the urban setting, the environmental and energy performance and the positive impact on social interactions and soft mobility.

Similarly, in DeCesaris's reflection in this volume, when she interprets the concept of 'green' at the zero level, she identifies its potential to rewrite the relationship between man and nature. The 'garden' at the neighbourhood scale, because of its minute dimension 'that allows it to fit with relative ease into the tissue of the urban fabric', can be configured as a new public space of proximity even in marginal, degraded, unused areas, through 'a way of relating to nature with informal, economic interventions', creating 'spaces for socialising with a value that is not only social but also therapeutic and productive'. On a territorial scale, the long-term and large-scale Carré de Soie project presented by Dupper shows the potential of the regeneration of this large urban area, linked to public transport systems and guaranteeing benefits for the local community. A community has championed the project and ensures its management, with the aim of keeping alive the local history linked to silk production and creating a green oasis in a dense urban structure, through a system of highly distinctive and participatory public spaces that 'invite social communication and enhance inclusion'.

11.4 Green Infrastructure for Sustainable Lifestyles and Green Economies

The arising attention to urban resilience from a multidisciplinary perspective has fostered the attention on the role of green systems as 'multifunctional landscapes' that provide ecological, socio-economic and cultural benefits towards a socio-ecological transition, combining the sustainability of territory and the promotion of eco-friendly and healthy lifestyles (Pickett et al. 2013). This multidisciplinary perspective requires overcoming the mere land use approach to green areas, as well as the concept of ecological network, promoting a holistic perspective and assessing the multiple benefits of nature for communities (MEA 2005) and individuals within the concept of 'green infrastructure'.

> Green infrastructure is the physical environment within and between our cities, towns and villages. It is a network of multi-functional open spaces, including formal parks, gardens, woodlands, green corridors, waterways, street trees and open countryside. It comprises all environmental resources, and thus a green infrastructure approach also contributes towards sustainable resource management (Town and Country Planning Association 2008).

At European level, green infrastructure is recognised as 'a strategically planned network of natural and semi-natural areas with other environmental features designed and managed to deliver a wide range of ecosystem services such as water purification, air quality, space for recreation and climate mitigation and adaptation. This network of green (land) and blue (water) spaces can improve environmental conditions and therefore citizens' health and quality of life. It also supports a green economy, creates job opportunities and enhances biodiversity' (EC 2013).

Therefore, this kind of infrastructure does not only perform ecological functions, but assumes complex meanings (Ravagnan and Poli 2017), related to 'full consideration of landscape diversity, attention to the richness and dissemination of cultural heritage and historical relationships networks, awareness of the density of economic, social and cultural dynamics that shape the territory' (Moccia 2010). In fact, the multifunctional use of green areas and water networks also enables a reconsideration of the ways and forms of collective re-appropriation and cultural production of open spaces, drawing new research and experimentation paths for the coordination of eco-friendly temporary uses and the 'construction of a multifunctional dimension of the spaces not to consume' (Gasparrini 2015).

In this perspective, the development of regeneration projects into resilient strategies (Masboungi 2005) can lead the city towards a new urban system that provides innovative social, cultural, economic and environmental answers to global changes, fostering new procedures able to renovate processes and to support multilevel governance and cooperative approaches. New partnership-based procedures highlight innovative experimentation paths combining the enhancement of natural and semi-natural open spaces and the participatory practices for eco-friendly cultural and leisure activities. These practices appear fundamental in particular in large urban and metropolitan areas where the pandemic has dramatically hit vulnerable groups, fostering socio-economic imbalances, isolation and health issues.

In this framework, the case study of the new special Collserola Range Protection Plan, illustrated by Eugènia Vidal-Casanovas, is an emblematic example of the role of green networks in metropolitan areas and a best practice for the synergies between thematic areas as well as the cooperation of different stakeholders. In particular, 'the wide variety of contributions of the team, within a multidisciplinary approach, have made it possible to define and develop an ecological strategy that covers various thematic areas, ranging from the preservation and improvement of ecological values to the enhancement of natural resources, public use, built heritage, infrastructures and services, and the functional space'. Furthermore, 'the systemic vision refers to the dialogue between different initiatives and stakeholders (...) moving from a position of public control to one of cooperation'.

11.5 Conclusions

The described approaches point out how the green networks design for sustainability and resilience require a combined approach.

On the one hand, a sustainable approach must pay attention to global challenges through the study and experimentation of common references and the standardisation of evaluation and monitoring in order to face together world issues such as climate change, pandemic and socio-economic inequalities.

On the other hand, resilience is related to specificity of ecological features and social practices, as well as their relationship, requiring local based approaches to territories and communities.

According to these approaches, therefore, the theme of the 'garden' goes beyond its archetypal concept, expanding in conceptual, theoretical, methodological and operational terms as a possible, and necessary, interpretative-planning and design component for intervening in the city and contemporary society.

In this way, outlining possible future paths of in-depth study, research and experimentation that adopt this approach and apply the references identified within the city's open laboratory.

Author Contributions Introduction and Conclusions by M.B. Andreucci (MBA), I. Poli (IP), C. Ravagnan (CR); Section 1 by MBA, Section 2 by IP, Section 3 by CR.

References

Andreucci MB, Loder A, Brown M, Brajković J (2021) Exploring challenges and opportunities of biophilic urban design: evidence from research and experimentation. Sustainability 13:4323. https://doi.org/10.3390/su13084323

Antonovsky A (1987) Unravelling the mystery of health. Jossey-Bass Inc, San Francisco

Beatley T (2016) Handbook of biophilic city planning and design. Island Press, Washington, DC

Beery T, Jönsson K, Elmberg J (2015) From environmental connectedness to sustainable futures: topophilia and human affiliation with nature. Sustainability 7(7):8837–8854. https://doi.org/10.3390/su7078837

Borja J, Muxi Z (2003) El espacio publico: ciudad y ciudadania. Electa, Barcelona

Bratman GN, Anderson CB, Berman MG, Cochran B, De Vries S, Flanders J, Folke C, Frumkin H, Gross JJ, Hartig T et al (2019) Nature and mental health: an ecosystem service perspective. Sci Adv 5. https://www.science.org/doi/10.1126/sciadv.aax0903

Braubach, M., Egorov, A., Mudu, P., Wolf, T., Thompson, C.W., & Martuzzi, M. (2017). Effects of urban green space on environ-mental health, equity and resilience. In N. Kabisch, H. Korn, J. Stadler, & A. Bonn (Eds.), Nature-based solutions to climate change adaptation in urban areas: linkages between science, policy and practice, pp. 187–205. Cham, Berlin/Heidelberg: Springer

EC (2013) Green infrastructure strategy. Green Infrastructure (GI) – enhancing Europe's natural capital. Bruxelles, SWD (2013) 155 final. Retrieved October 12, 2023, from https://environment.ec.europa.eu/topics/nature-and-biodiversity/green-infrastructure_en

Gambino R (2007) Patrimonio storico e paesaggio. In: Carullo S (ed) Spazi aperti nei contesti storici. Ancsa, Bergamo

Gasparrini C (2015) In the city on the cities. Nella città sulle città. List: Trento

Gehl J (2010) Cities for people. Island Press, Washington, DC

Hartig T, Mang M, Evans GW (1991) Restorative effects of natural environment experiences. Environ Behav 23:3–26. https://doi.org/10.1177/001391659123100

Kaplan S (1995) The restorative benefits of nature: toward an integrative framework. J Environ Psychol 15:169–182. https://doi.org/10.1016/0272-4944(95)90001-2

Kaplan R, Kaplan S (1989) The experience of nature: a psychological perspective. Cambridge University Press, Cambridge, UK

Kaplan R, Kaplan S (2005) Preference, restoration, and meaningful action in the context of nearby nature. In: Barlett P (ed) Urban place: reconnecting with the natural world. MIT Press, Cambridge, UK, p 330

Kellert SR, Wilson EO (1993) The biophilia hypothesis. Island Press, Washington, DC

Korpela KM, Ylén M, Tyrväinen L, Silvennoinen H (2008) Determinants of restorative experiences in everyday favorite places. Heal Place 14:636–652. https://doi.org/10.1016/j.healthplace.2007.10.008

Macchi Cassia C (1991) Il Grande Progetto Urbano. Roma, La Nuova Italia Scientifica

Masboungi A (2005) Régénérer les grands ensembles. Editions de La Villette, Paris

Marcus L, Colding J (2014) Toward an integrated theory of spatial morphology and resilient urban systems. Ecol Soc 19(4):55. https://doi.org/10.5751/ES-06939-190455

Masboungi A (2012) Projets urbains durables: stratégies. Le Moniteur, Paris

MEA (2005) A report of the millennium ecosystem assessment. Ecosystems and human Well-being. Island Press, Washington DC

Moccia FD (2010) Infrastruttura verde Urbanistica Informazioni 232:28–29

Pickett STA, Cadenasso ML, McGrath B (2013) Resilience in ecology and urban design: linking theory and practice for sustainable cities. Springer, New York/London

Ravagnan C, Poli I (2017) Green and blue networks: towards a safe future within risk management and strategic vision. Urbanistica 160:141–150

Secchi B (1989) Un progetto per l'urbanistica. Einaudi, Torino

Semapa (2010) L'îlot ouvert. Bruxelles, Christian de Portzamparc, AAM Editions

Town and Country Planning Association (2008) The essential role of green infrastructure: eco-towns green infrastructure worksheet, advice to promoters and Planners

Ulrich RS (1983) Aesthetic and affective response to natural environment. In: Altman I, Wohlwill JF (eds) Behaviour and the natural environment. Plenum Press, New York, pp 85–125

Ulrich RS (1993) Biophilia, biophobia, and natural landscapes. In: Kellert SEOW (ed) The biophilia hypothesis. Island Press, Washington, DC, pp 74–137

Index

F. Dal Falco (ed.), *Beyond the Garden*, Designing Environments,
https://doi.org/10.1007/978-3-031-50567-6

SPRINGER NATURE

GPSR Compliance

The European Union's (EU) General Product Safety Regulation (GPSR) is a set of rules that requires consumer products to be safe and our obligations to ensure this.

If you have any concerns about our products, you can contact us on ProductSafety@springernature.com

In case Publisher is established outside the EU, the EU authorized representative is:

Springer Nature Customer Service Center GmbH
Europaplatz 3
69115 Heidelberg, Germany

The manufacturer's authorised representative in the EU is Springer
Nature Customer Service Centre GmbH, Europaplatz 3, 69115 Heidelberg,
Germany. If you have any concerns regarding our products, please
contact ProductSafety@springernature.com

Printed and bound by CPI Group (UK) Ltd, Croydon, CR0 4YY
28/09/2026
02224077-0018